新航道学校指定SAT培训教材

Digital SAT

新SAT机考

数学金题精练

美国维布伦特 ◎ 编著

15年+经验的SAT团队编著

- 4 大考查领域
- 19 类细分考点
- 3 级难度划分
- 360+道练习题

紧扣官方考纲+全景式解读SAT机考新变革

正向推导，逆向排除
演算步骤环环相扣
逐步提升数学思维

世界知识出版社

图书在版编目（CIP）数据

新 SAT 机考数学金题精练 / 美国维布伦特编著 . -- 北京 : 世界知识出版社 , 2024.6
ISBN 978-7-5012-6765-1

Ⅰ. ①新… Ⅱ. ①美… Ⅲ. ①数学—高等学校—入学考试—美国—教学参考资料 Ⅳ. ①O1

中国国家版本馆 CIP 数据核字（2024）第 082453 号

图字：01-2024-2068

责任编辑 谢 晴
特约编辑 龚玲琳
特邀编辑 王丽娜 商 文
封面设计 韩 玥
责任出版 赵 玥
责任校对 张 琨

书 名 **新 SAT 机考数学金题精练**
Xin SAT Jikao Shuxue Jinti Jinglian

编 著 美国维布伦特

出版发行 世界知识出版社
地址邮编 北京市东城区干面胡同 51 号（100010）
网 址 www.ishizhi.cn
电 话 010-65233645（市场部）
经 销 新华书店
印 刷 清淞永业（天津）印刷有限公司
开本印张 889 毫米 × 1194 毫米 1/16 14 3/4 印张
字 数 280 千字
版次印次 2024 年 6 月第 1 版 2024 年 6 月第 1 次印刷
标准书号 ISBN 978-7-5012-6765-1
定 价 98.00 元

前言

PREFACE

自2023年春季伊始，SAT考试率先在国际考区迎来历史性变革——由传统纸笔考形式转型为机考模式。2024年春，北美考区紧随其后，全面推行机考改革。至此，SAT纸笔考正式退出历史舞台。机考的全面转型为广大学子带来前所未有的机遇与挑战。然而，备考资源之稀缺、结构改革之新颖，使率先踏入未知领域的中国考生成为首批探索者。为填补SAT机考改革后国内市场在备考资料领域的信息空白，新航道国际教育集团联合美国维布伦特出版社（Vibrant Publishers）倾力为中国SAT考生打造了“新SAT机考”系列图书。该系列图书由一支拥有超过15年SAT教学辅导经验的美国专家团队精心编著，旨在为2024年及之后参加SAT考试的中国考生提供全面且有针对性的机考练习资源。此系列图书自在亚马逊平台上线以来，便凭借其前沿的学术理念、实用的备考策略和卓越的品质保障，赢得了广大考生和教育工作者的广泛赞誉，成为当前市场上引领SAT机考备考新趋势的权威材料之一。

“新SAT机考”系列图书紧扣SAT机考的核心考点，通过精心设计的专项练习题和全真模拟卷，搭配深入浅出的答案解析，为考生提供了系统且全面的学习指导。此外，书中所有题目皆经过易、中、难的精准难度划分，可有效帮助考生在实践中评估自身知识掌握程度，从而在备考过程中更加游刃有余，轻松应战正式考试。

SAT机考变革最为显著的两大变化体现在考试结构的优化和多阶段自适应模式的引入。一方面，考试结构经过优化，将原本各自独立的阅读与文法部分合并为一个板块，同时考题数量也得到了合理精简：从154道题减至98道题，其中阅读与文法板块包含54道题目，数学板块包含44道题目。考生只需在2小时14分钟内即可完成整场考试，相较于传统纸笔考的3小时时长，考试效率得到显著提升。另一方面，多阶段自适应模式更是机考改革的一大亮点。该模式能够基于考生在考试中的实时表现，动态调整试题的难度，确保试题与考生实际能力相匹配。具体来讲，阅读与文法和数学部分均设有两个模块，每个模块包含一组试题。在模块一中，考生须面对不同难度的试题，随后系统会根据考生在这一模块中的表现，评估并确定模块二试题的难易程度。这种机制极大地提升了考试的公正性，可有效防止作弊等不端行为的发生。

在SAT机考的阅读与文法板块中，题干文本跨越了文学、科学、历史、社会研究和人文等多个学科领域，文本长度控制在25—150词之间。相较于传统纸笔考，机考试题文本虽短，但考查形式更为灵活多变，旨在通过一系列多元化的题型设计全面评估考生的阅读理解和语法能力。

“信息与观点”题型要求考生凭借深厚的知识储备和敏锐的分析技能，深入理解和阐释文章中明确或隐含的信息，同时借助相关图表精准地进行信息的识别、解释、评估和处理。

“技巧与结构”题型则着重考查考生在复杂语境中理解学术性词汇和短语的能力，以及从修辞学角度对文章进行准确评价，并在多篇主题相关的文章间建立联系的综合技能。

“观点表达”题型强调考生须根据特定的修辞目标，精准地运用专业知识，使书面表达更加行之有效。

“标准英语语法”题型要求考生展现其出色的编辑技能，确保书面表达完全符合标准英语句子结构和标点符号规范。

与传统纸笔考相比，SAT 机考取消了多个试题共用同一文本的长阅读形式，每个问题均配备独立文本。这一调整不仅要求考生具备更强的快速阅读和信息提取能力，并且对考生的逻辑判断能力提出了更高的要求。此外，阅读部分新增的诗歌载体，对于中国考生而言无疑是一个较大挑战。因此，考生在备考过程中需要系统学习这一新题型，以确保能够全面适应 SAT 机考所带来的全新要求和挑战。

在 SAT 机考的数学板块中，相较于传统纸笔考，其题型和考查形式基本维持原貌。不同之处在于考生现在可以在整个数学板块中自由使用计算器。（整个数学部分都有一个内置的图形计算器，学生也可以自带经批准使用的计算器。）此外，数学题干的文本篇幅亦有所缩减。此举旨在将考查重心回归于对数学概念的深入理解和公式的灵活应用，而非过度倚重计算速度与文本解读能力。

SAT 机考的数学核心考点依旧涵盖四大领域：

在代数部分，考生须展现扎实的分析能力与解题技巧，以准确构建并求解线性方程、不等式及方程组。

高等数学部分则深入检验考生对于复杂数学知识的掌握程度，考查考生对一元二次方程、多项式方程、绝对值方程、有理方程、非线性方程和根式方程的分析、求解与构建能力，以及对双变量线性与非线性方程的处理能力。

解决问题和数据分析部分着重考查考生对比率、速率和比例关系的理解，以及运用定量推理解决涉及单变量和双变量数据问题的技巧，以培养考生将数学理论知识应用于实际情境中的能力。

几何学和三角学部分要求考生具备出色的空间想象能力和几何推理能力，以解决与角度、周长、面积、体积、圆和三角形相关的复杂问题。这不仅是对考生几何知识的检验，更是对考生逻辑思维和数学技能应用水平的全面评估。

此外，SAT 机考在题型布局上进行了优化，将运算填写题穿插在单选题之中。这种混合题型的设计相较于纸笔考中两种题型明确分离的形式，更能全面考查考生在面对不同题型时的应变能力和综合解题技巧。

《新 SAT 机考数学金题精练》一书旨在为考生提供全面且详尽的备考指南，助力考生深入理解并熟练掌握数学的所有考点，成功突破高分壁垒。具体而言，本书的特色体现在以下几个方面：

一、机考改革，深度剖析新动向

本书开篇即对 SAT 机考改革进行了全面且周密的剖析，细致对比了纸笔考与机考在考试时长、题目数量、结构布局以及题型考点等方面的异同。此外，第一章还系统阐述了自适应模式的运作机制，并对机考中的常见问题进行了全面归纳。这些内容为考生深入理解并顺利适应机考的新特点和新变化提供了有力支持。

二、考点梳理，智慧引领实战路

第二章详细梳理了数学部分的整体框架及全部考点。每个考点都精心配置了针对性的样题、答案解析、解题技巧和备考建议。通过系统且全面的学习与练习，考生能够迅速理解并掌握SAT对考生能力的具体要求，进而提升解题能力与应试水平。

三、专项精练，答案详析助突破

第三章至第六章专注于四大核心考点——代数、高等数学、解决问题和数据分析、几何和三角学。每个考点均配备了大量习题及详尽的答案解析和干扰项分析。考生可根据自身薄弱环节，有针对性地进行强化训练，牢牢掌握出题模式和解题逻辑。在正式练习之前，建议考生根据个人情况，量身打造学习计划，确保每日都能完成既定数量的习题。不论做题结果正确与否，均请仔细阅读答案解析及干扰项分析，深入钻研答案背后的推理过程。此外，考生亦可先行挑选几个问题试答，然后核对答案，检查自己是否能够准确理解题干要求。如果遇到不解之处，请回头详阅本书对各类题型的系统阐释以及应对这些题型的实用技巧。

四、等级划分，难点攻克更有方

本书的答案解析对每道题目的难度和考点都进行了标注。通过比对答案，考生能够有效评估自身的知识掌握程度，进而确定后续备考的重点和方向。

五、全真模拟，评估水平展实力

本书末尾配置了一套数学全真模拟试题，共计44道题目，分设两个模块。此套试题严格遵循SAT实际考试标准，涵盖所有题型和难度梯度，旨在还原真实考试体验。通过完成此套模拟试题，考生可以全面检验自己的学习成果，自信迎接正式考试。

祝各位考生在SAT备考征程中一帆风顺，取得理想成绩！

编者

2024年4月

Table of Contents

Chapter 1

About the Digital SAT

Introduction

Now that you have made the important decision to head to college/university, there is one last thing you need to do to achieve your goal—taking the SAT. Most universities or colleges, including the Ivy League schools such as Yale, Harvard, and others expect you to have a good SAT score to secure admission in any course of your choice.

But, there is a major change in how students will give the SAT. The College Board has decided to transition the famous pencil-and-paper test into a fully digital one. The College Board's decision to go digital is based on giving a fair testing experience to students. The digital test will be easier to take, easier to administer, will be more secure, and more relevant.

For giving the new test, you need to be aware of the format of the test, the time that will be given to you to answer each question, the possible complexity of the questions, and the scoring method employed to assess your performance in the test. In this chapter, you will discover important information all that including the SAT policy of inclusive accessibility, the newly introduced Multistage Adaptive Testing feature, the modular format of the test, and much more.

The College Board has also streamlined the method of delivery of the digital SAT. With the latest test delivery platform for the Digital SAT Suite assessments, students can have access to all their tests and their content, as well as enjoy the chance of practicing with the full-length, adaptive practice test offered for free on the platform so that students can be aware of their knowledge levels before taking the real tests. More so, every question on the Digital SAT Suite is in a discrete (standalone) format. This indicates that test takers can answer each question independently. They don't necessarily need to refer to a common stimulus such as an extended passage.

If you are attempting the SAT for the first time, it could be scary not knowing exactly what to expect in the test. This is why this book is specifically designed to expose you to everything you need to know about successfully taking the Digital SAT Suite test.

Customized Test Delivery Platform

The College Board sets up a customized test delivery platform for the Digital SAT Suite assessments. This platform is designed according to the principles of UDA (Universal Design for Assessment) and the main goal of it is to make the testing experience accessible to maximum number of students. The most useful features of this platform are that: (i) all test takers can have complete access to the tests and their content; (ii) students will be able to take full-length, adaptive practice tests for free on the platform so that they can assess their knowledge levels or have an understanding of similar test materials before attempting the real tests.

Multistage Adaptive Testing

The College Board is changing from a linear testing mode, which has been the primary mode of SAT administration to an adaptive mode.

The main difference between the linear and adaptive testing modes is that for the linear testing mode, students are given a test form that contains some questions that have already been set before the test day and do not change during testing, irrespective of the student's performance.

On the other hand, the adaptive testing model makes it possible for the test delivery platform to adjust the questions' difficulty level based on the performance of the individual test takers. Therefore, each student will be given test questions that match their level of understanding.

This adaptive test mode used for the Digital SAT Suite is known as **Multistage Adaptive Testing (MST)**. The MST is administered in 2 stages, and each stage comprises a module or set of questions. The first module consists of test questions with different ranges of difficulty levels (easy, medium, and hard). The performance of the test takers in the first module is appropriately assessed, and the results are used to determine the level of difficulty of questions to be administered to them in the second module.

The set of an administered first-stage module and its second-stage module are referred to as a ***panel***.

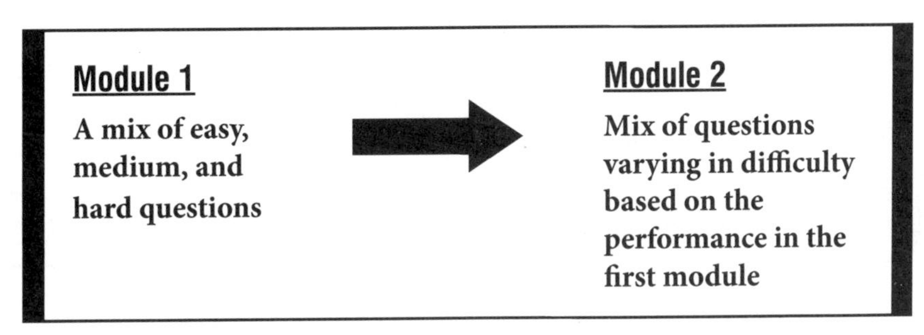

Embedded Pretesting

The Digital SAT Suite also includes embedded pretesting in its design. What this means is that a small number of pretest (unscored) questions are incorporated among the operational (scored) questions. Even though they are not administered for a score, students may not be able to distinguish these pretest questions from the operational questions on which their scores are based. It is advisable that students pay maximum attention and effort to these questions, which can be used in estimating their understanding levels to the difficulty of the questions. The number of pretest questions is few so you will not be asked to focus mainly on questions that won't be used to estimate your final SAT score. It is important to note that answers to pretest questions won't contribute to your final score. The pretest questions are mainly used to gather students' performance information so that it can be utilized later to assess if these questions are appropriate for operational use later.

Discrete Questions

One interesting aspect of the Digital SAT is that all their questions are in discreet format; that is they are standalone. You can answer each question on its own, which doesn't necessarily require any reference to a common stimulus such as an extended passage. This is one of the striking differences between the paper-and-pencil SAT and the Digital SAT in the sense that the former uses both discrete and question sets. In practice, the question sets expect you to reference a common stimulus.

Scoring

Students will obtain a section score based on their final performance on the Reading and Writing and Math section. For the SAT, students can get a score between 400–1,600. Hence, for each of the tests of the Digital SAT, there will be 3 scores reported: (1) A Reading and Writing section score; (2) A Math section score; (3) A total score, which is the sum of the two section scores. It is important to note that the scales for these scores have the same ranges as for the paper-based SAT Suite. This indicates that the Digital SAT total score is on the familiar 400–1,600 scale.

Reading and Writing	Math	Total Score
Between 200–800	Between 200–800	Between 400–1,600

Overall Test Specifications

The Digital SAT is made up of two sections: A Reading and Writing (RW) section and a Math section. In the linear model, the test has separate sections for Reading and Writing. However, in the Digital SAT, both the Reading and Writing tests are combined in one section. The questions in these two sections concentrate primarily on the skills and knowledge that students need to use in college and/or for getting ready for their careers. The main parts of the Digital SAT tests are similar to those of the paper-and-pencil SAT test assessments. More so, all the testing programs within the Digital SAT Suite, whether it is the SAT, PSAT 10, PSAT 8/9, or PSAT/NMSQT have similar designs. Although, these tests allow for differences in test takers' ages and levels of understanding.

Digital SAT Suite: Overall Test Specifications

Characteristic	Reading and Writing section	Math section
Administration	Two-stage adaptive design; this section contains two separately timed modules	Two-stage adaptive design; this section contains two separately timed modules
Number of questions	54 questions; 25 questions in each module with 2 pretest questions	44 questions; 20 questions in each module with 2 pretest questions
Time	64 minutes	70 minutes
Time per question	1.19 minutes	1.59 minutes
Time per module	32 minutes	35 minutes
Content domains	Information and Ideas, Craft and Structure, Expression of Ideas, Standard English Conventions	Algebra, Advanced Math, Problem-Solving and Data Analysis, Geometry and Trigonometry

Test Length

There are a total of 54 questions for the Reading and Writing section. These 54 questions are divided into two equal-length modules; that is, one for each of the section's two stages. Out of the 27 questions for each module, **25 questions are operational**—which means that test takers' performance on them is used to calculate their section score, and **2 questions are pretests**.

For the Math section, the first module has **20 operational questions** and **2 pretest questions**. Then the second module consists of **20 operational questions** and **2 pretest questions**. In total, the Math section will comprise 44 questions.

Time Per Module

You will have 32 minutes to complete each module of the Reading and Writing section and 35 minutes to complete each module of the Math section. Once the time for the first module has expired, test takers are automatically advanced to the second module. The second module may contain questions that are of higher or lower difficulty, depending on your performance in the first module. You will not have the opportunity to return to the first-module questions.

Total Time Allotted

You will have 64 minutes to complete the Reading and Writing section and 70 minutes to complete the Math section.

Average Time Per Question

You will, on average, have 1.19 minutes to answer each Reading and Writing question and 1.59 minutes to answer each Math question.

Question Format(s) Used

The Reading and Writing section mostly utilizes four-option multiple-choice questions, and each question has a single best answer **(which is referred to as the keyed response or key)**. Roughly 75 percent of questions in the Math section also adopt the same four-option multiple-choice format, while the remaining part of the test utilizes the **student-produced response (SPR) format.** This means that students will be required to answer the latter type of questions by giving their own responses and putting their responses in the field next to the question. These questions measure your ability to be able to solve math problems by yourself. It is possible for the SPR questions to have more than one correct response; however, you are required to provide only one answer.

Text Complexity

It is assumed that the complexity test takers can read is directly related to how ready they are for college and their careers. Therefore, the idea of text complexity is strictly considered when designing and developing the Digital SAT Suite. The texts in the Reading and Writing section are given three complexity bands (grades 6–8, grades 9–11, and grades 12–14).

Texts for grades 12–14 have the highest complexity, followed by the texts for grades 9–11, while grades 6–8 have texts with the lowest complexity. While it is possible to use the same texts for grades 12–14 and grades 9–11, those difficult texts cannot be used for grades 6–8 because they don't appropriately assess the literacy knowledge and skills of students in eighth and ninth grades.

On the other hand, text complexity is not an issue in the Math section, because it is not formally measured. It is estimated that about 70 percent of Math questions don't necessarily have a context. You are only required to use the information/data provided to solve some questions that may be related to science, social studies, history, etc.

What is Changing

The College Board continues to maintain fairness and high quality in its administration of SAT Suite, and some aspects of its operations are changing. These changes include:

- From 2024, students can no longer take the paper-and-pencil SAT tests.
- The Digital SAT Suite tests are particularly shorter than their paper-and-pencil predecessors—it can be taken in 2 hours 14 minutes instead of 3 hours.
- Test takers now have more time on their hands to answer each question.
- It is now possible for you to receive scores in days instead of weeks, faster than the predecessor paper-and-pencil SAT.
- The SAT Suite now connects students to opportunities based on their scores. They can be connected to information and resources concerning local 2-year colleges, career options, and workforce training programs.
- States, schools, and districts will be given more flexibility concerning when they could give the SAT Suite tests.
- The Digital SAT will now have a single Reading and Writing section instead of separate Reading and Writing and Language sections. More importantly, the Reading and Writing section's passages are significantly shorter and more diverse.
- A single (discrete) question is associated with each passage (or passage pair) instead of having several questions associated with a small number of longer passages, as it is for the paper-and-pencil SAT Suite tests.
- You can now use calculators throughout the Math section.

What is Staying the Same

Despite the above-mentioned changes, some aspects of the SAT Suite tests are remaining the same, such as:

- The Digital SAT will still measure skills and knowledge that you are learning in school that can be used in college and/or your future career.
- The test will be scored on the same scales as the paper-and-pencil test.
- The test will be administered in schools and test centers with a proctor.
- You will still be connected to scholarships and the College Board National Recognition Programs.
- Support will be given to all students who need accommodations and/or support to access the tests and their content.
- The Reading/Writing passages will cover a wide range of academic disciplines and text complexities.
- The test will still have both multiple-choice and student-produced response question formats.

More Questions Answered about the Digital SAT

1. **How will students take the Digital SAT?**

   You can give the SAT on a laptop or tablet using a custom-built digital exam application that can be downloaded in advance of the test day.

2. **How will the Digital SAT be more secure?**

   At this moment, if one test form is compromised, it can mean that the scores for all the students in that group or at the same test centers will be canceled. However, going digital will make it possible to give every student a unique test form so that it won't be technically possible to share answers.

3. **How will the College Board address test day issues and technical support challenges?**

   The College Board has dedicated customer service resources ready to troubleshoot issues on test day for students and test centers. There is a technology coordinator for each test center to provide additional support and technical help when needed.

4. **What kinds of tools will be available for students taking the Digital SAT?**

   You can use the following tools while using the software:

   - Flag questions to come back to them later
   - A countdown clock to know when you are running out of time. You can decide to show or hide it at the top of their testing screen
   - A built-in graphing calculator that you can use on the entire math section (or you can bring their own calculators)
   - A reference sheet, for each math question.

The Reading and Writing Section at a Glance

The table below summarizes the specifications of the types of questions and their distribution in the Reading and Writing section.

Content Domain	Skill/Knowledge	Question Distribution
Information and Ideas	• Central Ideas and Details • Command of Evidence ▪ Textual and Quantitative • Inferences	12–14 questions (26%)
Craft and Structure	• Words in Context • Text Structure and Purpose • Cross-Text Connections	13–15 questions (28%)
Expression of Ideas	• Rhetorical Synthesis • Transitions	8–12 questions (20%)
Standard English Conventions	• Boundaries • Form, Structure, and Sense	11–15 questions (26%)

Sample Questions

1. The term *"Anthropocene"* introduced by Dutch scientist Paul Crutzen in the mid-1970s, is often used in the context of pollution caused by human activity since the commencement of the Agricultural Revolution, but also pertains largely to all major human bearings on the environment.

   Various start dates for the Anthropocene have been offered by scientists, ranging from the beginning of the first Agricultural Revolution, also known as the *Neolithic Revolution*, to as recently as the 1960s. However, the _________ has not been completed, and hence, a formal, conclusive date remains to be finalized.

   Which choice completes the text with the most logical and precise word or phrase?

   A) ratification

   B) investigation

   C) legality

   D) approval

   Key: A

   Level: Hard | **Domain:** CRAFT AND STRUCTURE

   Skill/Knowledge: Words in context

   Key Explanation: Choice A is the correct option because "ratification" refers to the action of signing or giving formal consent to something, making it officially valid. This word is best suited to the context because the second paragraph of the passage talks about how many scientists have offered dates, but a conclusive date has yet to be finalized. The keywords to focus on are "formal, conclusive date" which points to which **Choice A** might be most suitable in this context.

Distractor Explanations: Choice B is incorrect because there is no evidence provided that an investigation may have been initiated into the subject. Similarly, **Choices C** and **D** are incorrect because the passage does not talk about any legalities or approval process that need to be completed for a date to be finalized.

2. Brazil's Atlantic Rainforest is among the most biodiverse regions in the world. But despite its spectacular diversity, ______________________. To counter this, the Society for the Conservation of Birds in Brazil advocates for birds in Brazil, their habitats, and biodiversity in general, and works towards sustainability in the use of natural resources. Their work focuses on educating local people on the importance of birds, biodiversity, and developing environmentally sustainable economic alternatives, along with good governance tools to empower local communities and improve the quality of life of local people.

   What choice most logically completes the underlined space?

   A) there are no more birds in the forest

   B) the Society for the Conservation of Birds in Brazil cannot do much

   C) the rainforest is under extreme threat from human development

   D) there are a number of steps that one can take to preserve the Atlantic Rainforest

   Key: C

   Level: Medium | **Domain:** INFORMATION AND IDEAS

   Skill/Knowledge: Command of evidence (textual)

   Key Explanation: Choice C is the best answer because the first sentence talks about the diversity of the Atlantic Rainforest, while the third sentence talks about what the Society for the Conservation of Birds in Brazil is doing to counter said problem in the second sentence. They work by "educating local people" and "developing environmentally sustainable economic alternatives." Therefore, it may be inferred that the problem denoted in the underlined portion of the text involves humans and economics. Hence, using the process of elimination, **Choice C** is the best answer.

   Distractor Explanations: Choice A is incorrect because the text mentions "conservation of birds in Brazil," which means that birds may be endangered but not extinct. **Choice B** is incorrect because there is no information provided that supports this statement. **Choice D** is incorrect because it does not fit in the context of the sentence.

The Math Section at a Glance

The table below summarizes the specifications of the types of questions and their distribution in the Math section.

Content Domain	Skill/Knowledge	Question Distribution
Algebra	• Linear equations in one variable • Linear equations in two variables • Linear functions • Systems of two linear equations in two variables • Linear inequalities in one or two variables	13–15 questions (35%)
Advanced Math	• Equivalent expressions • Nonlinear equations in one variable and systems of equations in two variables • Nonlinear functions	13–15 questions (35%)
Problem-Solving and Data Analysis	• Ratios, rates, proportional relationships, and units • Percentages • One-variable data: distributions and measures of center and spread • Two-variable data: models and scatterplots • Probability and conditional probability • Inference from sample statistics and margin of error • Evaluating statistical claims: observational studies and experiments	5–7 questions (15%)
Geometry and Trigonometry	• Area and volume • Lines, angles, and triangles • Right triangles and trigonometry • Circles	5–7 questions (15%)

Sample Questions

1. The dog park charges \$10 for a membership and \$3 per hour for the dog to run around in their park. Mindy brings her dog to the park and spends less than \$40. Which of the following inequalities represents Mindy's situation, where h is the number of hours at the park and C is the total amount Mindy paid?

   A) $3h + 10 < 40$

   B) $3C - 10 < 40$

   C) $3h + 10 = 40$

   D) $3h + 10 > 40$

   Key: A

   Level: Easy | **Domain:** ALGEBRA

   Skill/Knowledge: Linear inequalities in one or two variables | **Testing Point:** Creating a linear inequality

   Key Explanation: Choice A is correct. To determine the inequality that represents the situation, first create the expression that is equal to the total amount that Mindy paid (C).

   The total amount C is the sum of the membership fee (\$10) and the fee for having the dog in the park in h hours. This yields $C = 10 + 3h$ or $C = 3h + 10$.

   Since Mindy spent less than \$40 in the dog park, then $C < 40$. Substituting the value of C in terms of h in the inequality yields $3h + 10 < 40$.

   Therefore, the inequality $3h + 10 < 40$ is the correct answer.

   Distractor Explanations: Choice B is incorrect. This option is wrong because C is the total amount paid by Mindy and not the rate per hour for the dog to run around the park. **Choice C** is incorrect. Mindy spends less than \$40, and hence, it should be $<$ instead of $=$. **Choice D** is incorrect. Mindy spends less than \$40, and hence, it should be $<$ instead of $>$.

2. Which expression is equivalent to $2x^2 + 3x + 1$?

   A) $(2x + 1)(2x + 1)$

   B) $(x + 2)(x + 1)$

   C) $(x - 2)(x - 1)$

   D) $(2x + 1)(x + 1)$

   Key: D

   Level: Easy | **Domain:** ADVANCED MATH

   Skill/Knowledge: Equivalent expressions | **Testing Point:** Factoring a quadratic equation

   Key Explanation: Choice D is correct. To find the equivalent expression, factor the given quadratic equation by splitting the middle term.

In $2x^2 + 3x + 1$, $a = 2$, $b = 3$ and $c = 1$ using the format $ax^2 + bx + c$.

Getting the product of a and c yields $ac = (2)(1) = 2$.

The factors of 2 whose sum is the value of b (where $b = 3$) are 2 and 1.

Hence, the equation can be written as $2x^2 + 2x + x + 1$.

Grouping the binomials in the equation yields $(2x^2 + 2x) + (x + 1)$.

Factoring $2x$ from the first group yields $2x(x + 1) + (x + 1)$.

Factoring $x + 1$ from the two groups yields $(2x + 1)(x + 1)$.

Therefore, **Choice D** is the correct answer.

Distractor Explanations: Choice A is incorrect. It may result from a conceptual or calculation error. **Choice B** is incorrect. It may result from a conceptual or calculation error. **Choice C** is incorrect. It may result from a conceptual or calculation error.

Chapter 2

Overview of the SAT Math Section

Introduction

There are 44 questions in the entire Math section and they are divided into 2 modules of equal lengths, one representing each of the section's two stages. Hence, each module has 22 questions, of which 20 are operational and 2 are pretest. Only the answers to the operational questions count towards estimating your final Math score. The pretest questions are only used to collect data that will be used in judging whether such questions can be used in future tests.

Table 3.1 Digital SAT Math section specifications

Characteristic	Math Section
Mode of administering the test	The Math section is designed according to the multistage adaptive model and administered through two modules that are timed differently.
Time per module	1st module will take 35 minutes to complete. 2nd module will take 35 minutes to complete. The total time available for the section is 70 minutes.
Average time per question	Each question will take 1.59 minutes to be completed.
Score reported	You will be scored between 200–800. The score for this section represents half of the total score.
Question format used	The question format is discreet, with four multiple-choice options.
Informational graphics	Informational graphs can include line graphs, tables, and bar graphs.

Time Per Section and Module

You will be given 70 minutes to complete the Math Section. This indicates that you have a total of 35 minutes to complete each module. You will be automatically taken to the next module once the time for the first module has expired. You may be advanced to a higher or lower difficulty module based on your performance in the first module, and you won't be able to go back to the questions in the first module.

Average Time Per Question

It is estimated that you will have approximately 1.59 minutes per question.

Question Format Used

The SAT Math section utilizes two distinct question formats; they are the four-option multiple-choice format and the student-produced response or SPR format. For multiple-choice questions, you will be required to choose a single best answer known as the keyed response or key. This multiple-choice format accounts for 75 percent of the Math questions. However, for the SPR, you will do the calculations yourself and enter your correct response in the test platform. Although you may obtain different answers in SPR questions when you do the calculations yourself, you are encouraged to only enter the single, best/correct answer.

Context Topics

About 30 percent of all Math questions have context; that is, the questions are expressed in words using subject matters or topics that are obtained from social studies, real-world scenarios, or science. In addition to understanding the context, which you could be familiar with or can easily understand, you will also be required to use your knowledge of the context and Math skills to answer the questions. It is not expected that you should have specific knowledge of the topics beforehand.

Word Count by Question

For those Math questions expressed in words (context), most of them consist of 50 words or fewer.

Informational Graphics

Some of the Math questions contain informational graphics such as bar graphs, graphs of functions in the xy-plane, histograms, dot plots, scatterplots, line graphs, and representations of geometric figures. The informational graphics are included in the test questions for two main purposes: first, to indicate the importance of such graphics in Math to convey relevant information/data; second, to test your ability to solve mathematical problems by locating, interpreting, and utilizing the information displayed in the informational graphics.

Text Complexity

About 70 percent of Math questions don't have context (words). So, it is not possible to measure text complexity for those questions. However, for the remaining 30 percent of the Math questions with context drawn from social studies, science, and other real-world scenarios, efforts have been made to modify the text. In other words, the linguistic difficulty has been removed from such context so as to make them clear, direct, and simple for students to understand.

Domain Structure

The Math Section's questions fall under one of the four content domains described below:

- **Algebra:** You are expected to carefully analyze, properly solve, and create both linear equations and inequalities. More so, you will have to correctly analyze and solve different types of equations utilizing different methods.

- **Advanced Math:** You are required to show that you have the necessary knowledge and skills indicative of your progress toward advanced Math courses. As a matter of fact, you would need to demonstrate your ability to properly analyze, solve, interpret, and create appropriate equations that include but are not restricted to quadratic, polynomial, absolute value, rational, nonlinear, and radical equations. More so, you must analyze and solve both linear and nonlinear equations in two variables.

- **Problem-Solving and Data Analysis:** You are expected to make use of your understanding and quantitative reasoning concerning rates, proportional relationships, and ratios to analyze, interpret, and solve mathematical problems involving one-and-two-variable data.

- **Geometry and Trigonometry:** You are expected to utilize your problem-solving skills in solving questions relating to angles, perimeter, triangles, area, volume, circles, and trigonometry.

NOTE: Each question belongs to a single content domain, and the questions for each module are drawn from all four content domains. Therefore, in each domain, questions are designed to test some skills/knowledge points.

Tips for the Math Section

You would find these tips helpful when preparing for your Digital SAT Math test:

- Always read the SAT Math questions understandably before you start answering them. You don't want to waste your limited time providing answers to questions you vaguely understand.
- Pay attention to the choices for each question. Based on your prior knowledge of similar math questions in high school, you may be able to quickly eliminate some incorrect answers while looking for the most correct answer in the choices. This practice will also help you to save time that you can use for solving other difficult questions.
- If a math question seems very difficult for you on the first try, do not waste your time thinking about how you could solve it; move straight to the next question.
- Skip over questions that appear to be too wordy. You could come back later to attempt them after you have successfully completed the ones with shorter context.
- As calculators are now allowed in the Math section, it will help you arrive at the answers more quickly. However, you will still need to depend on your knowledge of arithmetic to successfully use it. Do not be overconfident about the benefits of using a calculator during your test because it may give you a wrong answer and a false hope. So, it is advisable that you use it responsibly and creatively in order to make sure that the answers you are getting are all correct.
- Getting a high score on the SAT math test depends on your preparations before the test. This is why it is very important that you try your hand at several practice math questions before taking the real Digital SAT test. You will be able to identify the most probable answers to some of the related questions you will be seeing while taking the real SAT test itself.
- Pay serious attention to numbers, mathematical signs, and expressions. If you make any mistake calculating with a wrong mathematical sign, for example, you can expect that your final answer will be wrong.
- In a situation in which you obtained two answers for the same question, you can quickly redo your calculations and eventually discover which is more correct. Enter ONLY one of the most appropriate answer for each question.

Algebra

For Algebra questions, you will be required to do some calculations to solve some mathematical problems relating to linear representations. In this way, you will be making sensible connections between different forms of linear equations. Your knowledge of high-school algebra lessons/topics will be very useful here.

Pay attention to the following skill/knowledge testing points for Algebra questions:

Skill/knowledge testing points

Linear equations in one variable

For example, $2x + 3 = 8$ is a typical one-variable linear equation having x as the single variable.

$x = 2.5$.

Sample Question

If $4x - (x + 6) = 3$, what is the value of 3^x?

A) 3

B) 9

C) 27

D) 81

Key: C

Level: Medium | **Domain:** ALGEBRA

Skill/Knowledge: Linear equations in one variable | **Testing Point:** Solving for x and using the value to solve for an exponential term

Key Explanation: Choice C is correct. Using distributive property yields $4x - x - 6 = 3$.

Adding 6 to both sides of the equation and combining like terms yields $3x = 9$.

Dividing both sides of the equation by 3 yields $x = 3$.

Hence, the value of 3^x is $3^3 = 27$.

Distractor Explanations: Choice A is incorrect. This is the value of x. **Choice B** is incorrect. This is the value of 3^2. **Choice D** is incorrect. This is the value of 3^4.

Linear equations in two variables

For example, $10x + 4y = 30$ is an example of a linear equation with two variables x and y.

Hence, if $y = 3$,

$$x = \frac{30-12}{10}$$

$x = 1.8$.

Sample Question

Line l, given by the equation $ax - 3y + 3 = 0$, is perpendicular to line k, given by the equation $-14x + by + 5 = 0$. What is the value of $\frac{a}{b}$?

A) $\frac{-14}{3}$

B) $\frac{-3}{14}$

C) $\frac{3}{14}$

D) $\frac{14}{3}$

Key: B

Level: Hard | **Domain:** ALGEBRA

Skill/Knowledge: Linear equations in two variables | **Testing Point:** Slope of perpendicular lines

Key Explanation: Choice B is correct. Lines that are perpendicular to each other have slopes that are negative reciprocals of the other. This can be represented as $m_1 \times m_2 = -1$. The slope of the first line can be given by $\frac{a}{3}$. The slope of the second line can be given by $\frac{14}{b}$. Therefore, $\frac{a}{3} \times \frac{14}{b} = -1$.

Multiplying both sides of the equation by $\frac{3}{14}$ yields $\frac{a}{b} = -\frac{3}{14}$.

Distractor Explanations: Choice A is incorrect and may result due to calculation or conceptual error. **Choice C** is incorrect and may result due to calculation or conceptual error. **Choice D** is incorrect and may result due to calculation or conceptual error.

Linear functions

A linear function forms a straight line on the graph.

For the function of x, $f(x) = 3x - 2$,

If $x = 4$.

Then, $f(x) = 12 - 2$.

Therefore, $f(x) = 10$.

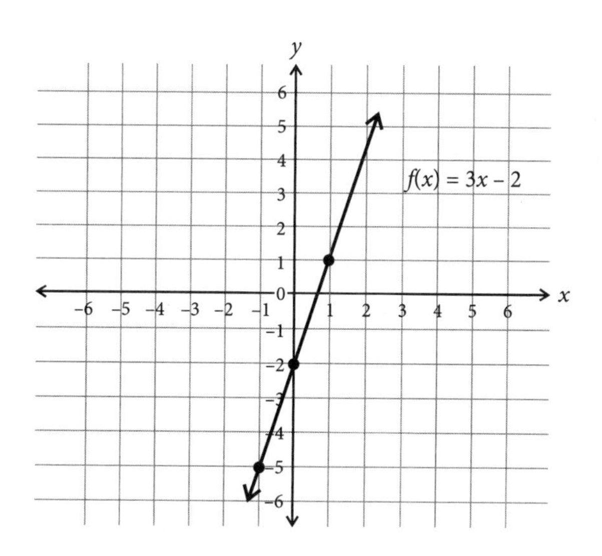

If plotted on a graph, $f(x)$ will appear like a straight line shown above.

Sample Question

If $g(x) = f(x + 2) + 2$ and $f(x) = 4x + 2$, what is the value of the y-intercept of $g(x)$?

A) 2

B) 8

C) 10

D) 12

Key: D

Level: Medium | **Domain:** ALGEBRA

Skill/Knowledge: Linear functions | **Testing Point:** Function translation

Key Explanation: Choice D is correct. To determine the y-intercept of $g(x)$, evaluate $g(x) = f(x + 2) + 2$.

Solving $f(x + 2)$ yields $f(x + 2) = 4(x + 2) + 2$.

Using distributive property yields $f(x + 2) = 4x + 8 + 2$.

Simplifying the equation yields $f(x + 2) = 4x + 10$.

Substituting the equivalent expression for $f(x + 2)$ in $g(x)$ yields $g(x) = 4x + 10 + 2$.

Simplifying the equation yields $g(x) = 4x + 12$.

Therefore, the y-intercept is 12.

Distractor Explanations: Choice A is incorrect. This is the y-intercept of the function $f(x)$. **Choice B** is incorrect and may be due to miscalculation or conceptual error. **Choice C** is incorrect and may be due to miscalculation or conceptual error.

Systems of two linear equations in two variables

If you are given two linear equations $2x + y = 8$ and $3x + 2y = 12$, you should solve the first equation for y, which will give you $y = -2x + 8$. Then, substitute for y in the second equation.

You will have $3x + 2(-2x + 8) = 12$

$3x - 4x + 16 = 12$

$-x = -4$.

Multiply both sides of the equation by -1

$x = 4$.

Sample Question

$$2y = px + 6$$

$$3y = 2x + 9$$

What is the value of p for the system of equation above if the system has no solution?

A) -3

B) $\frac{4}{3}$

C) 2

D) 3

Key: B

Level: Medium | **Domain:** ALGEBRA

Skill/Knowledge: Systems of two linear equations in two variables | **Testing Point:** Solving for linear systems that have no solutions

Key Explanation: Choice B is correct. For linear systems to have no solutions, the linear equations have to be parallel to each other. Converting $2y = px + 6$ to slope-intercept form yields $y = \frac{p}{2}x + 3$. The slope of the equation is therefore $\frac{p}{2}$.

Converting $3y = 2x + 9$ to slope-intercept form yields $y = \frac{2}{3}x + 3$. The slope of the equation is therefore $\frac{2}{3}$.

Since the lines must be parallel, their slopes must be the same.

Equating the two slopes yields $\frac{p}{2} = \frac{2}{3}$.

Multiplying both sides of the equation by 2 yields $p = \frac{4}{3}$.

Distractor Explanations: Choice A is incorrect. This option would result in system of perpendicular linear equations and would have one solution. **Choice C** is incorrect. This option would result to a system of equation with one solution as the linear equations would have different slopes. **Choice D** is incorrect. This option would result to a system of equation with one solution as the linear equations would have different slopes.

Linear inequalities in one or two variables

This is an example of a linear inequality equation with one variable, x: $ax + b < c$ where a, b, and c are real numbers.

Hence, if $a = 1$, $b = 4$, and $c = 10$, $x < 10 - \frac{4}{1}$.

$x < 6$.

However, for a linear inequality equation in two variables, $ax + by < c$, where a, b, and c are real numbers, and b is not equal to 0.

Hence, if $a = 2$, $b = 3$, and $c = 20$, $y = 1$, x will be $x < \frac{20 - 3}{2}$.

$x < 8.5$.

Sample Question

Carly wants to buy oranges and apples. Each orange costs \$0.40 and each apple costs \$0.35. If she can spend no more than \$6.90, what is the maximum number of oranges she can get if she buys at least 2 apples?

A) 13

B) 14

C) 15

D) 16

Key: C

Level: Medium | **Domain:** ALGEBRA

Skill/Knowledge: Linear inequalities in one or two variables | **Testing Point:** Solving linear inequalities

Key Explanation: Choice C is correct. Since Carly buys at least 2 apples, she can buy the maximum number of oranges by buying exactly 2 apples.

Hence, the problem can be represented by $2(0.35) + x(0.4) \leq 6.9$ where x is the maximum number of oranges that Carly can buy.

Simplifying the inequality yields $0.7 + 0.4x \leq 6.9$.

Subtracting 0.7 to both sides of the inequality yields $0.4x \leq 6.2$.

Dividing both sides of the inequality by 0.4 yields $x \leq 15.5$.

The maximum number of oranges would therefore be 15.

Distractor Explanations: Choice A is incorrect and may result from a conceptual or calculation error. **Choice B** is incorrect and may result from a conceptual or calculation error. **Choice D** is incorrect and may result from a conceptual or calculation error.

PEMDAS Rule: The PEMDAS Rule is a mathematical rule that helps students to know how to solve a math question that involves different mathematical signs at the same time in a math question.

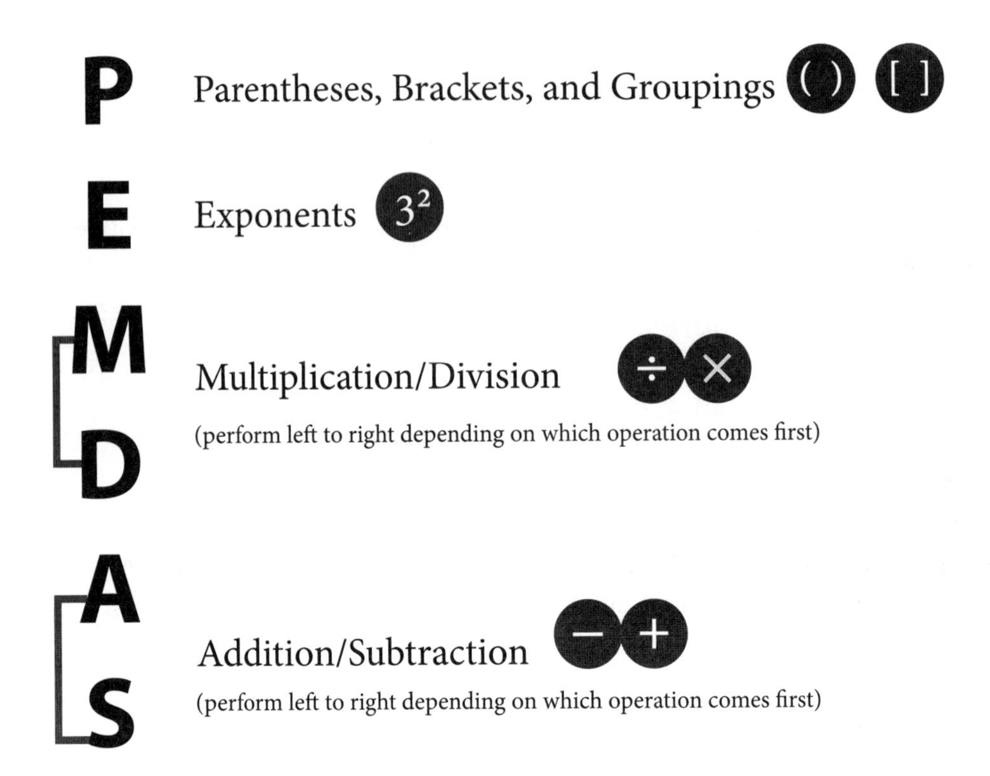

For example, when you have a math problem with different signs shown above, you will start first with the parentheses, brackets, and groupings, followed by exponents, then you will do the multiplication/division, and finally addition/subtraction. When solving linear equations, functions, and inequalities, it is very important to pay attention to the mathematical signs in them.

Advanced Math

For Advanced Math, you will be expected to interpret, calculate, correctly solve, and make use of structure, and create quadratic equations, absolute value, radical equations, exponential, rational, polynomial, and other related nonlinear equations. More so, you will be required to make appropriate connections between different forms of a nonlinear relationship between two variables.

For Advanced Math questions, these are the skill/knowledge testing points:

Skill/knowledge testing points

Equivalent expressions

Equivalent expressions are defined as expressions that will have the same value, even if they look different if the same values are used to substitute the variables in the expressions.

For example,

Do you think that $5x + 2$ and $5(x - 1) + 7$ are equivalent expressions?

Yes, they are!

NOTE: When solving for equivalent expressions, use the following ideas:

- Discover the coefficient in the expressions $a(bx + c) = abx + ac$, a, b, and c are the coefficients
- Combine the variables, which could be x or y
- Solve for the unknown variable, this could be x or y
- Rearrange the final formula

Sample Question

If $2^{\frac{3}{2}}$ is equivalent to $\sqrt[n]{4^3}$, what is the value of n?

A) 1

B) 2

C) 3

D) 4

Key: D

Level: Medium | **Domain:** ADVANCED MATH

Skill/Knowledge: Equivalent expressions | **Testing Point:** Exponential expressions

Key Explanation: Choice D is correct. If $2^{\frac{3}{2}} = \sqrt[n]{4^3}$ and $\sqrt[n]{4^3}$ is equivalent to $4^{\frac{3}{n}}$, then $2^{\frac{3}{2}} = 4^{\frac{3}{n}}$. Making the base of $4^{\frac{3}{n}}$ to 2 yields $2^{2\left(\frac{3}{n}\right)}$. Hence, $2^{\frac{3}{2}} = 2^{2\left(\frac{3}{n}\right)}$.

Equating the exponents yields $\frac{3}{2} = 2\left(\frac{3}{n}\right)$.

Simplifying the equation yields $\frac{3}{2} = \frac{6}{n}$.

Therefore, n is equal to 4.

Distractor Explanations: Choice A is incorrect and may be due to conceptual or calculation error. **Choice B** is incorrect and may result from a conceptual or calculation error. **Choice C** is incorrect and may result from a conceptual or calculation error.

Nonlinear equations in one variable and systems of equations in two variables

When *nonlinear equations* are plotted on the graphs, they give curves or nonlinear representations on the coordinate axis, while *linear equations* give straight lines (see the diagram below).

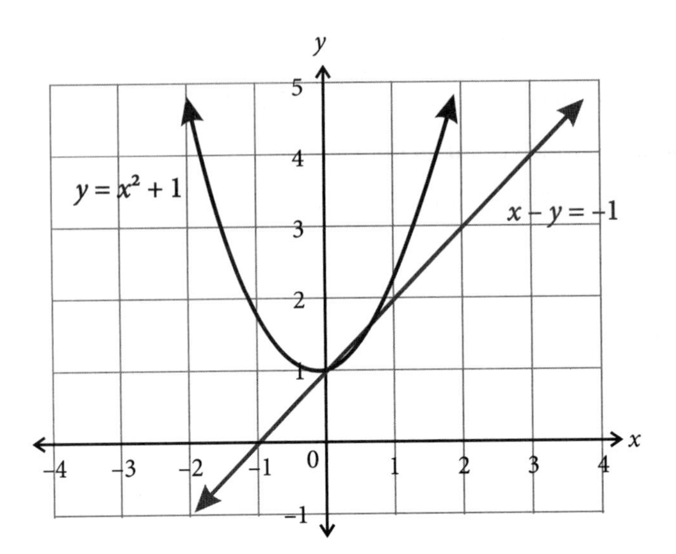

Nonlinear equations can have either one variable (x or y) or two variables (x and y).

There are three approaches that can be used to solve nonlinear equations:

- **By using a graph:** After obtaining real numbers for each of the variables, you can then plot y against x.
- **By substitution:** You can substitute the variables with the given or calculated numbers.

For example, in these equations, solve for y:

$$x^2 - y = 0$$

$$y = x - 2$$

To solve this, this equation solves for y, $y = x - 2$.

Then, solve for x by substituting y in the equation $x^2 - y = 0$.

$x^2 - x + 2 = 0$ (a quadratic equation).

You can use a quadratic formula to solve for x.

$x = \dfrac{-b \pm \sqrt{b^2 - 4ac}}{2a}$.

- **By elimination:** You can eliminate each variable from the equation using the real values (numbers) obtained from your calculations.

Sample Question

$$3(x - 3)(x - 6) = (x - 4)(x - 3)$$

How many solutions does the system above have?

A) Zero

B) One

C) Two

D) Infinitely many

Key: C

Level: Hard | **Domain:** ADVANCED MATH

Skill/Knowledge: Nonlinear equations in one variable and systems of equations in two variables | **Testing Point:** Finding the number of solutions for a system of nonlinear equations

Key Explanation: Choice C is correct. To find the number of solutions to the equation above, use the distributive property which yields $3(x^2 - 3x - 6x + 18) = (x^2 - 4x - 3x + 12)$.

Simplifying the equation yields $3x^2 - 27x + 54 = x^2 - 7x + 12$.

Subtracting x^2 and 12 and adding $7x$ to both sides of the equation yields $2x^2 - 20x + 42= 0$.

To find the number of solutions in a quadratic equation, use the discriminant $b^2 - 4ac$. The value of $a = 2$, $b = -20$, and $c = 42$.

Solving the value of discriminant yields $(-20)^2 - 4(2)(42)$; $400 - 336 = 64$.

Since the discriminant is greater than 0, the number of solutions for the system of equations is two.

Distractor Explanations: Choice A is incorrect and may result from a conceptual or calculation error. **Choice B** is incorrect and may result from a conceptual or calculation error. **Choice D** is incorrect and may result from a conceptual or calculation error.

Nonlinear functions

As its name implies, nonlinear functions are not linear, and they are not in the form of $f(x) = ax + b$.

NOTE: If you are not fast at plotting graphs to solve linear equation problems, concentrate only on using substitution and elimination methods because you don't have a lot of time to waste struggling with graph-plotting.

Some examples of nonlinear functions are:

- **Quadratic function:** $f(x) = x^2$
- **Exponential function:** $f(x) = 2^x$
- **Cubic function:** $f(x) = x^3 - 3x$

Sample Question

The population of bacteria in a pond is 3,030 and the population triples every 32 days. Which of the following functions best represents the population of the bacteria $P(t)$ in the pond after t days?

A) $P(t) = 3{,}030(3)^{\frac{t}{32}}$

B) $P(t) = 3{,}030(3)^{\frac{32}{t}}$

C) $P(t) = 3{,}030(3)^{3t}$

D) $P(t) = 3{,}030(3)^{32t}$

Key: A

Level: Medium | **Domain:** ADVANCED MATH

Skill/Knowledge: Nonlinear functions | **Testing Point:** Exponential growth function

Key Explanation: Choice A is correct. The exponential growth function has a formula of $f(x) = a(1 + r)^x$ where a is the initial amount, $(1 + r)$ is the growth rate and x is the time interval.

Since the initial number of bacteria in the pond is 3,030 then $a = 3{,}030$.

Since the population triples then $(1 + r) = 3$.

Since the population triples every 32 days then $x = \frac{t}{32}$.

Therefore, the equation for the function is $P(t) = 3{,}030(3)^{\frac{t}{32}}$.

Distractor Explanations: Choice B is incorrect and may result from using a wrong time interval. **Choice C** is incorrect and may result from using a wrong time interval. **Choice D** is incorrect and may result from using a wrong growth rate.

Problem-Solving and Data Analysis

In this question domain, you will be utilizing your quantitative reasoning to solve math questions relating to rates, units, percentages, and proportional relationships. More so, you will be required to properly analyze and correctly interpret some data distributions and then use that data to identify related probabilities, frequency, and conditional probabilities. You will also need to use fit models to discover and compare instances of exponential and linear growth.

You may be asked to do some calculations on the mean, median, and range of different values, including their standard deviations. You will be expected to have a basic grasp of study design and detect any margin of error in the data while doing your calculations.

Most of the skill/knowledge and testing points highlighted below may have been taught in your high school, and they will be tested by SAT for your college or career requirements.

Skill/knowledge testing points

Ratios, rates, proportional relationships, and units

The questions in this skill/knowledge will ask you to solve questions using proportional relationship between quantities, calculating or using a ratio or rate, and or using units, derived units, and unit conversion.

Sample Question

A cube has a volume of 96 cm^3. A larger version of this cube is made with twice the side lengths of the original. What is the volume of the larger cube?

A) 192

B) 288

C) 384

D) 768

Key: D

Level: Hard | **Domain:** PROBLEM-SOLVING AND DATA ANALYSIS

Skill/Knowledge: Ratios, rates, proportional relationships, and units | **Testing Point:** Ratios in similar figures

Key Explanation: Choice D is correct. The two cubes are similar and therefore the ratio of their side length is given by 1:2. The ratio of their volumes will be $1^3:2^3$ which is equivalent to 1:8. Therefore, the volume of the larger cube will be $8 \times 96 = 768$.

Distractor Explanations: Choice A is incorrect and may result due to calculation or conceptual error. **Choice B** is incorrect and may result due to calculation or conceptual error. **Choice C** is incorrect and may result due to calculation or conceptual error.

Percentages

Familiarize yourself with how to do simple calculations on unit percentages. Most importantly, memorize the necessary steps in achieving correct answers while attempting questions on these math topics. So, diligently do your practice by laying your hands on several examples relating to these testing points.

Sample Question

The price of a book at a bookstore is $45. Maria bought the book at a discounted price of $40. What is the value of the discount percentage?

A) 1.11%

B) 5%

C) 11.11%

D) 12.5%

Key: C

Level: Easy | **Domain:** PROBLEM-SOLVING AND DATA ANALYSIS

Skill/Knowledge: Percentages | **Testing Point:** Discounts and percentages

Key Explanation: Choice C is correct. Percentage discount is found by $\frac{discount}{original\ price} \times 100\%$. The discount is $45 – $40 = $5. The original price is $45. Therefore, the discount percentage is $\frac{5}{45} \times 100\% = 11.11\%$.

Distractor Explanations: Choice A is incorrect and might result when it is multiplied by 10% or due to mistake in placing decimals. **Choice B** is incorrect. This is the value of discount. **Choice D** is incorrect and may be due to conceptual or calculation error.

One-variable data: distributions and measures of center and spread

In math, you can do the distribution and measuring of center and spread in data sets by calculating their mean, median, range, deviation, etc. Revise the formulas for calculating the mean, median, range, and deviation of data sets.

Sample Question

Temperatures in City *X* and City *Y* were recorded on eight different days. The temperatures were then plotted on a scatterplot shown below. What is the range in temperature for City *Y*?

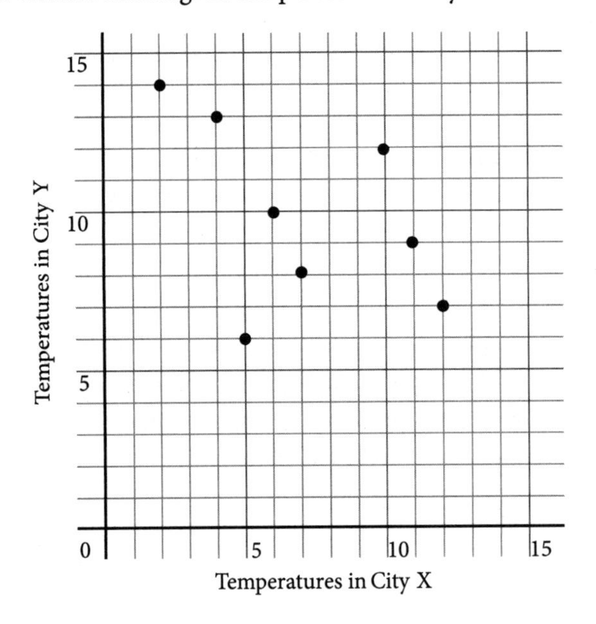

A) 8

B) 10

C) 12

D) 14

Key: A

Level: Medium | **Domain:** PROBLEM-SOLVING AND DATA ANALYSIS

Skill/Knowledge: One-variable data: distributions and measures of center and spread | **Testing Point:** Finding range using a scatter plot

Key Explanation: Choice A is correct. The range in temperature in City *Y* can be found by the difference between the maximum and the minimum temperature in City *Y*. The maximum temperature is 14 and the minimum temperature is 6. Therefore, the difference is 8.

Distractor Explanations: Choice B is incorrect and shows the value of range of temperature in City X. **Choice C** is incorrect and gives the maximum temperature in City X. **Choice D** is incorrect and shows the maximum temperature of City Y.

Two-variable data: models and scatterplots

In math, we use models and scatterplots to understudy and measure the distribution of two-variable data. The scatterplots can show clearly that a data set has: (i) a positive correlation; (ii) a negative correlation; or (iii) no correlation. Concerning the models, a data set may be of: (i) a linear model; (ii) a quadratic model; or (iii) an exponential model.

Sample Question

A meteorologist measures the amount of rain in inches on 7 consecutive days. On which days is the average rate of change greatest?

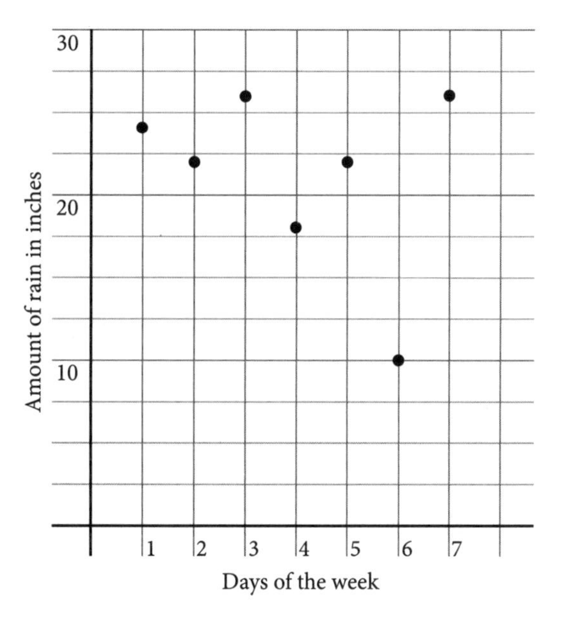

A) Day 3 and Day 6

B) Day 6 and Day 7

C) Day 2 and Day 3

D) Day 4 and Day 7

Key: B

Level: Hard | **Domain:** PROBLEM-SOLVING AND DATA ANALYSIS

Skill/Knowledge: Two-variable data: models and scatterplots | **Testing Point:** Average rate of change on a scatterplot

Key Explanation: Choice B is correct. The average rate of change between points is similar to the slope between the points. Using the points (6, 10) and (7, 26), the slope is $\frac{26-10}{7-6}=16$. This is the greatest average rate of change in the days on the scatterplot.

Distractor Explanations: Choice A is incorrect. This would result in a negative rate of change. **Choice C** is incorrect. This option may result due to conceptual or calculation error. **Choice D** is incorrect. This option may result due to conceptual or calculation error.

Probability and conditional probability

For probability questions, you will have to use one-and-two-way tables, area models, and other representations to find the relative frequency, probabilities, and conditional probabilities.

Sample Question

If a twelve-sided dice numbered from 1 to 12 is rolled, what is the probability that it lands on a multiple of 3?

A) $\frac{1}{12}$

B) $\frac{1}{6}$

C) $\frac{1}{4}$

D) $\frac{1}{3}$

Key: D

Level: Medium | **Domain:** PROBLEM-SOLVING AND DATA ANALYSIS

Skill/Knowledge: Probability and conditional probability | **Testing Point:** Finding the probability of an event

Key Explanation: Choice D is correct. Probability is found by $\frac{\textit{Favourable Outcomes}}{\textit{Total Outcomes}}$. The multiples of 3 in a dice of 12 are 3, 6, 9 and 12. The number of favorable outcomes is 4. The total number of outcomes is 12. Therefore, the probability is $\frac{4}{12}$, which is $\frac{1}{3}$.

Distractor Explanations: Choice A is incorrect and is the probability of the die landing in any number. **Choice B** is incorrect and may be due to conceptual error. **Choice C** is incorrect and may be due to conceptual error.

Inference from sample statistics and margin of error

In this case, you will be asked to make inferences by analyzing some data sets and using that information to answer some questions.

Sample Question

A researcher is conducting a study on the weight of students in a particular school. He randomly chooses 30 students and finds that their mean weight is 53 kg. He concludes that the margin of error in relation to the study is 3.5 kg. Which of the following statements can best be concluded about the weight of the students?

A) The 30 students in the survey all weigh between 49.5 kg and 56.5 kg.

B) The range weight of the students is 7kg.

C) It is plausible that the median of the 30 students in the survey lies between 49.5 kg and 56.5 kg.

D) It is plausible that the mean of the 30 students in the survey lies between 49.5 kg and 56.5 kg.

Key: D

Level: Medium | **Domain:** PROBLEM-SOLVING AND DATA ANALYSIS

Skill/Knowledge: Inference from sample statistics and margin of error | **Testing Point:** Discounts and percentages

Key Explanation: Choice D is correct. The margin of error implies that the plausible value of the mean would be 53 kg ± 3.5 kg. This means that the plausible value of the mean lies in the range of 49.5 kg and 56.5 kg. This statement would therefore be true.

Distractor Explanations: Choice A is incorrect and is false based on the information given in the study. **Choice B** is incorrect and is false based on the information given in the study. **Choice C** is incorrect and is false based on the information given in the study.

Evaluating statistical claims: observational studies and experiments

To evaluate statistical claims, you will not be required to do any calculations. All you need to do is to:

(i) Recognize good and bad sampling methods used in collecting the data sets.

(ii) Make your convincing conclusion based on the results of the surveys and experiments used in obtaining the data sets.

Sample Question

A researcher conducted research on different types of tea and its effect on the quality of sleep. The teas are chamomile, hibiscus, and peppermint. A random sample of 30 university students was selected for the study and asked to drink one kind of tea before sleeping. One-third of the sample drank chamomile tea while half of the remaining students drank hibiscus tea. Based on the experiment, the researcher concluded that chamomile tea is twice as effective as hibiscus tea and peppermint tea is half as effective as hibiscus tea. If 60% of students that drank hibiscus tea reported not having slept better, what is the total percentage of the sample that have slept better?

A) 20%

B) 40%

C) 47%

D) 53%

Key: C

Level: Hard | **Domain:** PROBLEM-SOLVING AND DATA ANALYSIS

Skill/Knowledge: Evaluating statistical claims: observational studies and experiments | **Testing Point:** Evaluating conclusions from experimental studies

Key Explanation: Choice C is correct. Since $\frac{1}{3}$ of the sample drank chamomile tea, then $\frac{1}{3} \times 30 = 10$ students drank chamomile tea before sleeping.

Since half of the remaining students drank hibiscus tea, then $\frac{30-10}{2}$ students drank hibiscus tea.

This means that 10 students also drank peppermint tea.

Since 60% of students who drank hibiscus tea reported not having slept better, then 40% reported to have slept better. This means that $10(1-0.6) = 4$ students slept better.

Since the researcher concluded that chamomile tea is twice as effective as hibiscus tea, then $40\% \times 2 = 80\%$ of the students or 8 students have slept better.

Since peppermint tea is half as effective as hibiscus tea, then the effectivity of peppermint is $\frac{40\%}{2} = 20\%$. This means that 2 students slept better when they drank peppermint tea.

The total number of students who slept better is $4 + 8 + 2 = 14$.

Therefore, the total percentage of the sample that have slept better is $\frac{14}{30} \times 100\% = 47\%$.

Distractor Explanations: Choice A is incorrect and may result from calculating the percentage of students that drank peppermint who have better sleep. **Choice B** is incorrect and may result from calculating the percentage of students that drank hibiscus who have better sleep. **Choice D** is incorrect and may result from calculating the total percentage of the sample that did not have better sleep.

Geometry and Trigonometry

Geometry and Trigonometry questions require you to solve mathematical problems related to area, volume, length, and scale factors concerning geometric figures. You will also need to use concepts and theories to identify similarity, congruence, and sufficiency in vertical angles, parallel lines that are cut by a transversal, and triangles.

You will be asked to solve math problems by utilizing the Pythagorean theorem, and your knowledge of right triangle and circle trigonometry. Do not forget the properties of certain types of right triangles. More so, you will need to use some properties and theorems relating to circles to solve problems. You will have learned all about these math topics in high school, and SAT will test you on them so as to meet the requirements for your college admission or future career.

Spend some time to practice any math questions you can lay your hands on that are related to these skill/knowledge testing points:

Skill/knowledge testing points

Area and volume

You will have to find the area, perimeter, volume, surface area of a geometric figure by using the given information.

Sample Question

A cone has the same volume as a cube. If the cone has a volume of 729 cm^3, what is the area of one surface on the cube?

A) 9

B) 36

C) 81

D) 486

Key: C

Level: Medium | **Domain:** GEOMETRY AND TRIGONOMETRY

Skill/Knowledge: Area and volume | **Testing Point:** Volume of a cube

Key Explanation: Choice C is correct. The volume of the cone is equal to that of the cube and is therefore 729 cm^3. The volume of the cube is found by $s^3 = 729$.

Hence, the value of $s = \sqrt[3]{729} = 9\,cm$.

The area of one surface of the cube can be found by $s^2 = 9^2 = 81\ cm^2$.

Distractor Explanations: Choice A is incorrect. This gives the value of one side of the cube. **Choice B** is incorrect. This gives the perimeter of one side of the cube. **Choice D** is incorrect. This gives the surface area of cube with a side length of 9 cm.

Lines, angles, and triangles

Do you remember your high-school math lessons on lines, angles, and triangles? Some questions you will meet in the Digital SAT math are related to lines, angles, and triangles. Here, you will solve questions relating to the congruence and the similarity of triangles.

Sample Question

The sum of interior angles in a regular 6-sided figure is $a\pi$ in radians. What is the value a?

A) 2

B) 4

C) 6

D) 8

Key: B

Level: Hard | **Domain:** GEOMETRY AND TRIGONOMETRY

Skill/Knowledge: Lines, angles, and triangles | **Testing Point:** Converting angles to degrees

Key Explanation: Choice B is correct. The sum of interior angles of a regular polygon can be given by $180(n - 2)$ where n is the number of sides. Therefore, the sum of interior angles of a 6-sided figure can be given by $180(6 - 2) = 720°$. To convert this to radians, multiply by $\frac{\pi}{180}$. This yields $720\times\frac{\pi}{180}=4\pi$. Therefore, $a = 4$.

Distractor Explanations: Choice A is incorrect and may result due to conceptual or calculation error. **Choice C** is incorrect and may result due to conceptual or calculation error. **Choice D** is incorrect and may result due to conceptual or calculation error.

Right triangles and trigonometry

You will need to look at the methods for solving right triangles' questions, which may include their area, perimeter, and semi-perimeter. More so, revise the different approaches for doing trigonometry calculations, estimating their hypotenuse, adjacent, and opposite. Don't forget to also revise sine, cosine, tangent, cosecant, cotangent, and secant functions.

This trigonometry table will help you to solve your trigonometry questions:

TRIGONOMETRY TABLE

	0°	30°	45°	60°	90°	120°	180°	270°	360°
sin α	0	$\frac{1}{2}$	$\frac{\sqrt{2}}{2}$	$\frac{\sqrt{3}}{2}$	1	$\frac{\sqrt{3}}{2}$	0	-1	0
cos α	1	$\frac{\sqrt{3}}{2}$	$\frac{\sqrt{2}}{2}$	$\frac{1}{2}$	0	$-\frac{1}{2}$	-1	0	1
tan α	0	$\frac{1}{\sqrt{3}}$	1	$\sqrt{3}$	∞	$-\sqrt{3}$	0	∞	0
cot α	∞	$\sqrt{3}$	1	$\frac{1}{\sqrt{3}}$	0	$\frac{1}{\sqrt{3}}$	∞	0	∞
sec α	1	$\frac{2}{\sqrt{3}}$	$\sqrt{2}$	2	∞	-2	-1	∞	1
cosec α	∞	2	$\sqrt{2}$	$\frac{2}{\sqrt{3}}$	1	$\frac{2}{\sqrt{3}}$	∞	-1	∞

Sample Question

Triangles ABE and ACD are right triangles. The length of $AB = 3\ cm$, $BC = 6\ cm$, and $AE = 5\ cm$. What is $cos\ \angle CDA$?

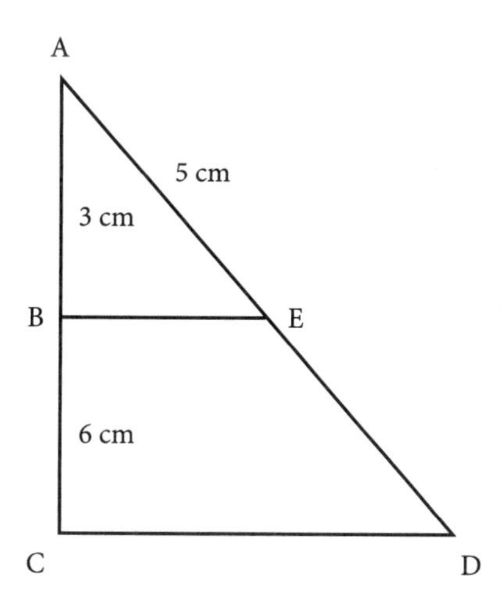

A) $\frac{3}{5}$

B) $\frac{4}{5}$

C) $3\left(\frac{3}{5}\right)$

D) $3\left(\frac{4}{5}\right)$

Key: B

Level: Medium | **Domain:** GEOMETRY AND TRIGONOMETRY

Skill/Knowledge: Right triangles and trigonometry | **Testing Point:** Trigonometry in similar triangles

Key Explanation: Choice B is correct. Since triangle ABE and triangle ACD are similar triangles, then $cos\ \angle CDA$ is equivalent to $cos\ \angle AEB$.

$cos\angle\ AEB$ is adjacent/hypotenuse which is $\frac{BE}{AE}$.

The length of BE can be found using the Pythagoras' theorem ($a^2 + b^2 = c^2$) which yields $3^2 + b^2 = 5^2$.

Simplifying the equation yields $b = \sqrt{25-9} = \sqrt{16} = 4$.

Hence, $BE = 4\ cm$.

Therefore, $cos\ \angle AEB = cos\ \angle CDA = \frac{4}{5}$.

Distractor Explanations: Choice A is incorrect. This option would give the value of $sin\ \angle CDA$. **Choice C** is incorrect and may result due to conceptual or calculation error. **Choice D** is incorrect and may result due to conceptual or calculation error.

Circles

Find the arc lengths, sector areas, and angles in circles using properties, definitions, and theorems related to circles.

Sample Question

The equation of a circle is represented by $x^2 + 6x + y^2 - 8y = 24$. Which of the following represents the coordinates to the center of a circle if it is translated to the left by 3 units?

A) (−3, 4)

B) (3, −4)

C) (0, −4)

D) (−6, 4)

Key: D

Level: Hard | **Domain:** GEOMETRY AND TRIGONOMETRY

Skill/Knowledge: Circles | **Testing Point:** Finding the center of a circle

Key Explanation: Choice D is correct. First, transform the given equation to standard form by completing the perfect square trinomials.

Adding 9 and 16 to both sides of the equation and grouping the trinomials yields $(x^2 + 6x + 9) + (y^2 - 8y + 16) = 24 + 16 + 9$.

Factoring the perfect square trinomials and simplifying the right side of the equation yields $(x + 3)^2 + (y - 4)^2 = 49$.

Hence, the center of the circle is (−3, 4).

Therefore, if this coordinate is translated 3 units to the left, then the new coordinate is (−6, 4).

Distractor Explanations: Choice A is incorrect. This is the original center of the circle before translation. **Choice B** is incorrect. This option may be due to conceptual or calculation errors. **Choice C** is incorrect. This option is due to translating the equation of the circle 3 units to the right.

Chapter 3

Algebra

This chapter includes questions on the following topics:

- Linear equations in one variable
- Linear equations in two variables
- Linear functions
- Systems of two linear equations in two variables
- Linear inequalities in one or two variables

ALGEBRA

1

Irene and Tabitha go out to lunch at a local restaurant. Irene's lunch costs $\$c$ and Tabitha's lunch costs \$2 more than Irene's. If they split the bill evenly and both paid a 20% tip, which expression below represents the amount of money that Irene spent?

A) $1.2(c + 1)$

B) $1.2(2c)$

C) $0.2(c + 1)$

D) $1.2(2c + 2)$

2

When Jen is 15 years old, her brother's age can be calculated using the expression $3x - 12$ where x is Jen's age at that time. What is the difference between Jen's and her brother's ages?

3

In the xy-plane, the graph of which of the following equations is a line with a slope of -4?

A) $y = \frac{1}{4}x$

B) $y = x - 4$

C) $y = -4x + 7$

D) $y = -4 + 4x$

4

When 3 times the number m is added to 12, the result is 33. What number results when 15 times m is added to 9?

5

Three cars have traveled 336 miles in total. Car A traveled 3 times as far as Car B, and Car C traveled twice as far as Car B. How many miles did Car C travel?

6

When $\frac{1}{3}$ is divided by the reciprocal of a particular number, the result is 16 more than the number.

What is that number?

7

What is the value of x in the equation below?

$$3x + 18 = 27$$

8

Tyrone is creating a study schedule for his midterms for the whole week including weekends. If he wants to study for at least 50 hours per week, with 10 hours per day over the weekend, then how many hours does he need to study per weekday?

A) 3

B) 4

C) 5

D) 6

1. **Level:** Hard | **Skill/Knowledge:** Linear equations in one variable | **Testing Point:** Converting English to Algebra with percentages

   Key Explanation: Choice A is correct. In order to determine which expression represents the situation, first solve the total cost of their foods. Adding the prices of Irene and Tabitha's lunches yields $c + (c + 2)$ or $2c + 2$. If they split the bill in half, then each of them needs to pay $\frac{2c+2}{2}$ or $c + 1$ for the food only. If both of them gave a 20% tip, then multiply the expression by 1.2. Therefore, the expression $1.2\,(c + 1)$ represents the total amount of money that Irene spent.

   Distractor Explanations: Choice B is incorrect and may result from not considering that Irene and Tabitha split the bill evenly. **Choice C** is incorrect and may result from calculating only the tip that each of them gave. **Choice D** is incorrect and may result from calculating the total amount of money they spent together.

2. **Level:** Easy | **Skill/Knowledge:** Linear equations in one variable | **Testing Point:** Interpreting word problem to solve a linear equation

   Key Explanation: The answer is 18. Calculate the brother's age by substituting Jen's age with the given expression which yields $3(15) - 12$ or 33 years old. Therefore, the difference between Jen's age (15 years old) and her brother's age is $33 - 15 = 18$.

3. **Level:** Easy | **Skill/Knowledge:** Linear equations in one variable | **Testing Point:** Determining the slope from the slope-intercept equation of a line

   Key Explanation: Choice C is correct. The slope-intercept form of a linear equation is written as $y = mx + b$ where m is the slope and b is the y-intercept. The slope is -4, so m is -4. The only answer choice with a slope of -4 is **Choice C**.

   Distractor Explanations: Choice A is incorrect and reflects error in identifying the slope in a linear equation. **Choice B** is incorrect and reflects error in identifying the slope in a linear equation. **Choice D** is incorrect and reflects error in identifying the slope in a linear equation.

4. **Level:** Medium | **Skill/Knowledge:** Linear equations in one variable | **Testing Point:** Solving a linear equation in one variable

   Key Explanation: The correct answer is 114. An equation can be set up and solved with the first sentence which yields $3m + 12 = 33$.

   Subtracting both sides of the equation by 12 yields $3m = 21$.

   Dividing both sides of the equation by 3 yields $m = 7$.

   The second sentence can be converted into the expression $15m + 9$. Substituting the value of m to the expression yields $15(7) + 9 = 114$.

5. **Level:** Medium | **Skill/Knowledge:** Linear equations in one variable | **Testing Point:** Solving a linear equation

   Key Explanation: The correct answer is 112 miles. Let the variables A, B and C be the distance traveled by Car A, Car B, and Car C, respectively. Since they traveled a total of 336 miles, then $A + B + C = 336$. Since Car A traveled 3 times as far as Car B, then $A = 3B$. Since Car C traveled twice as far as Car B, then $C = 2B$. Substituting the values of A and C in terms of B to the first equation yields $3B + B + 2B = 336$. Combining like terms yields $6B = 336$. Dividing both sides of the equation by 6 yields $B = 56$. Since $C = 2B$, then $C = 2(56) = 112$. Therefore, Car C traveled 112 miles.

6. **Level:** Medium | **Skill/Knowledge:** Linear equations in one variable | **Testing Point:** Creating a linear equation and solving

   Key Explanation: –24 is correct. Let x be the unknown number. Therefore, the equation will be $\frac{1}{3} \div \frac{1}{x} = x + 16$.

   Simplifying the left side of the equation yields $\frac{x}{3} = x + 16$.

   Multiplying 3 to both sides of the equation yields $x = 3x + 48$.

   Subtracting $3x$ from both sides of the equation yields $-2x = 48$.

   Dividing both sides of the equation by –2 yields $x = -24$.

7. **Level:** Easy | **Skill/Knowledge:** Linear equations in one variable | **Testing Point:** Solving for x in a linear equation

   Key Explanation: 3 is correct. To solve for the value of x, subtract 18 from both sides. This yields $3x + 18 - 18 = 27 - 18$ or $3x = 9$. Divide both sides of the equation by 3. The result would be $x = 3$.

8. **Level:** Medium | **Skill/Knowledge:** Linear equations in one variable | **Testing Point:** Creating and solving an linear equation in one variable

   Key Explanation: Choice D is correct. Begin by creating an equation to represent the total number of hours Tyrone needs to study in terms of x hours per 5 weekdays, if he studies 10 hours per day over the two weekend days:
   $50 = 20 + 5x$
   $30 = 5x$
   $x = 6$ hours per day.

   Distractor Explanations: Choice A is incorrect and reflects errors in interpreting the value of variables in the word problem. **Choice B** is incorrect and reflects errors in interpreting the value of variables in the word problem. **Choice C** is incorrect and reflects errors in interpreting the value of variables in the word problem.

9

When $x = 21$, what is the value of y in the function $21x + 21 = \frac{35y}{2}$?

A) $\frac{33}{5}$

B) 21

C) $\frac{132}{5}$

D) 330

10

When $\frac{2}{3}x + 4 = 10y$, and $x = 3$, what is the value of y?

A) $\frac{3}{5}$

B) $\frac{10}{13}$

C) $\frac{18}{5}$

D) 5

11

I. $-2x + y = 3$

II. $2y - x = 12$

III. $y = -\frac{1}{2}x + 7$

IV. $y = -2x - 7$

Which one of the following statements is true based on the four linear equations above?

A) Lines I and II are perpendicular.

B) Lines I and IV are parallel.

C) Lines I and III are perpendicular.

D) Lines II and III are parallel.

12

A farmer started with 8 rows of corn. He planned to add 3 rows during the summer months of June, July, and August for the next 7 years. If an equation is written in the form $y = mx + b$ to represent the number of rows of corn y the farmer planted x years after his first summer crop, what is the value of b?

13

$$b - \frac{1}{4}c = 0$$

If $b = 2$ in the equation above, what is the value of c?

Linear equations in two variables

14

Leslie is a technician for a cable company. Each day, she receives multiple locations with cable boxes that need repair. The number of boxes left that she still needs to fix at the end of each hour can be estimated with the equation $C = 12 - 8h$ where C is the number of cable boxes left and h is the number of hours worked that day. What is the meaning of the value 12 in this equation?

A) Leslie will complete the repairs within 12 hours.

B) Leslie starts each day with 12 boxes to fix.

C) Leslie repairs cable boxes at a rate of 12 boxes per hour.

D) Leslie repairs cable boxes at a rate of 12 per hour.

15

$$y = 28.2 - 2.5x$$

What are the coordinates of the y-intercept of the above equation?

A) (0, 28.2)

B) (2.5, 0)

C) (0, 2.5)

D) (28.2, 0)

16

Students in a geography bee score 10 points for every correctly answered question and lose 5 points for every incorrectly answered question. If a student answers 14 questions correctly but misses the final question, what is his final score?

17

Jenna deposits an initial amount of money into an account with simple interest. If the interest rate is 3% per year and she originally deposited $1,000, which equation below represents the value ($y) of her account after t years?

A) $y = 30t + 1{,}000$

B) $\frac{y}{1{,}000} = 30t$

C) $y = 1{,}000 - 30t$

D) $y = 1{,}000t - 30$

18

What is the value of the x-intercept of a line that is parallel to $6x - 2y = 8$ and passes through the point (5, 9) ?

A) -6

B) 2

C) 3

D) 4

19

Sally sold f fries and b burgers at her snack bar this weekend. If the price for fries is \$3.50 and the price for a burger is \$5.00, and there is no sales tax or tip, which of the following represents the total amount of money Sally made at her snack bar this weekend?

A) $5fb + 3.50$

B) $5b + 3.50f$

C) $3.50b + 5f$

D) $3.50fb + 5$

20

If y is 2 more than 3 times x and $x = \frac{1}{2}$, what is the sum of $x + y$?

A) $\frac{1}{2}$

B) 2

C) $\frac{7}{2}$

D) 4

21

The formula to convert degrees Celsius to degrees Fahrenheit can be represented by the equation $\frac{9}{5}C + 32 = F$, where C is the temperature in degrees Celsius and F is the temperature in degrees Fahrenheit.

Which of the following equations represents C in terms of F?

A) $C = \frac{9}{5}(F - 32)$

B) $C = \frac{5}{9}(F - 32)$

C) $C = \frac{9}{5}(F + 32)$

D) $C = \frac{5}{9}(F + 32)$

22

The formula $\frac{9}{5}C + 32 = F$ is used to convert the temperature from degrees Celsius to degrees Fahrenheit. If the temperature increases by two degrees Celsius, what would the equivalent increase be in degrees Fahrenheit? Round your answer to the nearest tenth.

A) 1.1

B) 2

C) 3.6

D) 35.6

23

A line passes through points $\left(2, \frac{1}{2}\right)$ and (4, 2).

What is the equation of a line perpendicular to this line, which passes through (3, –2)?

A) $y = -\frac{4}{3}x - 2$

B) $y = \frac{3}{4}x + 2$

C) $y = \frac{3}{4}x - 1$

D) $y = -\frac{4}{3}x + 2$

24

A furniture store is having a sale on lamps and rugs. Which function below represents the total cost of purchase (T) after the 20% discount of (l) lamps and (r) rugs if both cost $75 each?

A) $T = 0.2\ (75rl)$

B) $T = 0.8\ (75r + 75l)$

C) $T = 1.2\ (75r + l)$

D) $T = 0.2\ (l + 75r)$

25

If $5y + 7y + 15y + 42y = 3y + 14y + z$, what is the value of y?

A) $y = \frac{52}{z}$

B) $y = \frac{z}{52}$

C) $y = \frac{1}{51}$

D) $y = 51$

26

If $\frac{a-1}{2} = b$ and $b = 2$, what is the value of $a + 1$?

A) 4

B) 5

C) 6

D) 7

27

If $6m - n = 24$, what is $\frac{64^{2m}}{4^n}$?

A) 2^{64}

B) 4^{24}

C) 64^{6}

D) The value cannot be determined from the information given.

28

$$y = 22 + 3.9x$$

One end of a spring is attached to the top of the tent. When an object of mass x grams is attached to the other end of the spring, the spring stretches to a length of y inches as shown in the equation above. What is x when y is 69?

A) 47

B) 12.05

C) 17.69

D) 23.33

29

If $x = \frac{7y-4}{4}$, what is the value of y in terms of x?

A) $\frac{x}{y}$

B) $\frac{4x}{7}$

C) $\frac{x+1}{7}$

D) $\frac{4x+4}{7}$

9. **Level:** Easy | **Skill/Knowledge:** Linear equations in two variables | **Testing Point:** Evaluating a linear equation given an input value

Key Explanation: Choice C is correct. Substituting 21 for x in the equation yields $21(21)+21=\frac{35y}{2}$. Simplifying the left side of the equation by multiplication and then addition yields $462=\frac{35y}{2}$. Multiplying both sides of the equation by 2 gives $924 = 35y$.

Then dividing both sides of the equation by 35 gives $\frac{132}{5}=y$ or $y=\frac{132}{5}$.

Distractor Explanations: Choice A is incorrect and may result from dividing 462 by 2, instead of multiplying by 2. **Choice B** is incorrect and may result from a conceptual or calculation error. **Choice D** is incorrect and may result from a conceptual or calculation error.

10. **Level:** Easy | **Skill/Knowledge:** Linear equations in two variables | **Testing Point:** Solving linear equations in two variables for a value of one of the variables

Key Explanation: Choice A is correct. When $x = 3$, substitute 3 for x in the equation $\frac{2}{3}x+4=10y$ to get $\frac{2}{3}(3)+4=10y$. Simplifying the left side of the equation by multiplication and then addition, yields $6 = 10y$. Dividing both sides by 10 gives $y=\frac{6}{10}=\frac{3}{5}$.

Distractor Explanations: Choice B is incorrect and may result from inverting the fraction $\frac{6}{10}$. **Choice C** is incorrect and may result from adding the values of x and y. **Choice D** is incorrect and may result from dividing the value of x by the value of y.

11. **Level:** Easy | **Skill/Knowledge:** Linear equations in two variables | **Testing Point:** Determining whether lines are perpendicular or parallel

Key Explanation: Choice C is correct. To compare the slopes of the four lines given, the equations should be in the slope-intercept form of a line $y = mx + b$, where m is the slope and b is the y-intercept of the line. Lines III and IV are already in the slope-intercept form. To get Line I into slope-intercept form add $2x$ to both sides of the equation to get $y = 2x + 3$. To get Line II into slope-intercept form add x to both sides of the equation and then divide all terms of the equation by 2 to get $y = \frac{1}{2}x + 6$. Parallel lines have the same slope and different y-intercepts and perpendicular lines have opposite sign reciprocal slopes. Lines I and III are perpendicular because they have opposite sign reciprocal slopes. The slope for Line I, is -2 and for Line III it is $\frac{1}{2}$. This makes **Choice C** the correct answer.

Distractor Explanations: Choice A is incorrect. It would be selected because of not understanding correctly the relationships between parallel and perpendicular lines and slope. **Choice B** is incorrect. It would be selected because of not understanding correctly the relationships between parallel and perpendicular lines and slope. **Choice D** is incorrect. It would be selected because of not understanding correctly the relationships between parallel and perpendicular lines and slope.

12. **Level:** Medium | **Skill/Knowledge:** Linear equations in two variables | **Testing Point:** Converting English to Algebra and solving a linear equation

    Key Explanation: The correct answer is 8. The slope-intercept form of a linear equation is $y = mx + b$ where m is the slope, and b is the y-intercept or initial y value. The problem states that y represents the number of rows of corn and that the farmer started with 8 rows of corn. Therefore, $b = 8$.

13. **Level:** Easy | **Skill/Knowledge:** Linear equations in two variables | **Testing Point:** Solving a linear equation for one variable with the given value of the other variable

    Key Explanation: The correct answer is 8. Substituting 2 for b in the given equation yields $2-\frac{1}{4}c=0$. Subtracting 2 from both sides of the equation results in $-\frac{1}{4}c=-2$. Multiplying both sides of the equation by -4 yields $c = 8$.

14. **Level:** Medium | **Skill/Knowledge:** Linear equations in two variables | **Testing Point:** Determining the meaning in context of a term in a linear equation

    Key Explanation: Choice B is correct. At the start of the day when $h = 0$, the value of C is equal to 12. Therefore, Leslie starts each day with 12 cable boxes to fix since C is the number of boxes left and h is the number of hours worked that day.

    Distractor Explanations: Choice A is incorrect and reflects error in interpreting functions in word problems. **Choice C** is incorrect and reflects error in interpreting functions in word problems. **Choice D** is incorrect and reflects error in interpreting functions in word problems.

15. **Level:** Easy | **Skill/Knowledge:** Linear equations in two variables | **Testing Point:** Finding the y-intercept of a linear equation

    Key Explanation: Choice A is correct. The y-intercept is the y-value where x is equal to 0. Substitute 0 for x and solve y which yields $y = 28.2 - 2.5(0)$.

    Simplifying the equation yields $y = 28.2$ when $x = 0$. Therefore, the coordinates are (0, 28.2).

    Distractor Explanations: Choice B is incorrect and reflects error in interpreting linear functions. **Choice C** is incorrect and reflects error in interpreting linear functions. **Choice D** is incorrect and reflects error in interpreting linear functions.

16. **Level:** Easy | **Skill/Knowledge:** Linear equations in two variables | **Testing Point:** Creating and using a linear equation in two variables

    Key Explanation: 135 is correct. Begin by creating an expression to represent the situation of x, correct answers, and y, incorrect answers which yields $10x - 5y$.

    Then substitute the number of correctly and incorrectly answered questions which yields $10(14) - 5(1) = 140 - 5 = 135$.

17. **Level:** Easy | **Skill/Knowledge:** Linear equations in two variables | **Testing Point:** Creating a linear equation from information in question

    Key Explanation: Choice A is correct. In order to create an equation that represents the linear relationship, first identify the amount that the deposit increases per year, or the slope of the linear equation. Since the interest rate is 3% of the original deposit, the interest per year is $1,000(0.03) = \$30$. Hence, the interest after t years is $30t$. Since her initial deposit is \$1,000, the correct equation is $y = 30t + 1,000$.

Distractor Explanations: Choice B is incorrect and is likely the result of an incorrect combination of variables. **Choice C** is incorrect and is likely the result of an incorrect combination of variables. **Choice D** is incorrect and is likely the result of an incorrect combination of variables.

18. **Level:** Easy | **Skill/Knowledge:** Linear equations in two variables | **Testing Point:** Finding the x-intercept with parallel lines

Key Explanation: Choice B is correct. The given equation of the line is $6x - 2y = 8$. The slope of a line can only be found when the equation is in the slope-intercept form which would be $y = 3x - 4$. The slope m is equal to 3. The parallel line would also have a slope of 3 which makes its equation $y = 3x + b$ where b is the y-intercept. Plugging in the point (5, 9) to get the value of b yields $9 = 3(5) + b$. Simplifying the equation yields $b = -6$. Hence, the equation of the parallel line is $y = 3x - 6$. The x-intercept is the value of x when y is equal to 0. Substituting 0 to y yields $3x - 6 = 0$. Simplifying the equation yields $x = 2$. The x-intercept would be 2.

Distractor Explanations: Choice A is incorrect and is the y-intercept of the parallel line to a given equation. **Choice C** is incorrect and is the slope of the lines. **Choice D** is incorrect and is the value of y-intercept of the given equation.

19. **Level:** Medium | **Skill/Knowledge:** Linear equations in two variables | **Testing Point:** Constructing a linear equation given the variables and their coefficients

Key Explanation: Choice B is correct. To create an expression that represents the total amount of money Sally made, the variables, which represent the number of fries and burgers sold, must be matched and multiplied with the corresponding prices. The number of fries sold (f), is multiplied with the price \$3.50, and the number of burgers sold (b), is multiplied with the price \$5. The total amount of money Sally made can be calculated by adding the two products. Therefore, the expression must be $5b + 3.50f$.

Distractor Explanations: Choice A is incorrect because the number of fries and burgers sold are multiplied by the wrong price. **Choice C** is incorrect as the prices of fries and burgers is interchanged. **Choice D** is incorrect as the number of fries and burgers sold are multiplied by the wrong price.

20. **Level:** Easy | **Skill/Knowledge:** Linear equations in two variables | **Testing Point:** Converting English to Algebra and solving a linear equation for the sum of variables given one variable

Key Explanation: Choice D is correct. Translating the words into equation form yields $y = 3x + 2$. Substituting the value of x yields $y = 3\left(\frac{1}{2}\right) + 2$ or $y = \frac{7}{2}$. Adding the values of x and y yields $\frac{1}{2} + \frac{7}{2}$ or $\frac{8}{2}$. Therefore, $x + y = 4$.

Distractor Explanations: **Choice A** is incorrect because it gives value of x. **Choice B** is incorrect and may result from a conceptual or calculation error. **Choice C** is incorrect because it gives value of y.

21. **Level:** Easy | **Skill/Knowledge:** Linear equations in two variables | **Testing Point:** Solving for one variable in terms of another

Key Explanation: Choice B is correct. In order to express the relationship in terms of F, rearrange the equation to equal to C. Subtracting 32 from both sides of the equation yields $\frac{9}{5}C = F - 32$. Multiplying 5 and dividing 9 by from both sides of the equation yields $C = \frac{5}{9}(F - 32)$.

Distractor Explanations: Choice A is incorrect and may result from a conceptual or calculation error. **Choice C** is incorrect and may result from a conceptual or calculation error. **Choice D** is incorrect and may result from a conceptual or calculation error.

22. **Level:** Easy | **Skill/Knowledge:** Linear equations in two variables | **Testing Point:** Solving for one variable in terms of another

Key Explanation: Choice C is correct. Modify the equation in order to express an increase in Celsius by two degrees. Therefore, the new temperature in degrees Fahrenheit is $F_{new} = \frac{9}{5}(C+2)+32$. Getting the difference between the new and original equation yields $F_{new} - F = \left[\frac{9}{5}(C+2)+32\right] - \left(\frac{9}{5}C+32\right)$. Using the distributive property and combining like terms to simplify the equation yields $F_{new} - F = \frac{9}{5}(C+2) - \frac{9}{5}C$. Factoring out $\frac{9}{5}$ yields $\frac{9}{5}(C+2-C)$ or $\frac{9}{5}(2)$. Therefore, the equivalent increase in degrees Fahrenheit is $\frac{18}{5}$ or 3.6.

Distractor Explanations: Choice A is incorrect and most likely features arithmetic errors in calculating the change in degree in Fahrenheit. **Choice B** is incorrect and may result from an error in applying the distributive property. **Choice D** is incorrect and most likely features arithmetic errors in calculating the change in degree in Fahrenheit.

23. **Level:** Medium | **Skill/Knowledge:** Linear equations in two variables | **Testing Point:** Finding the equation of a line given two points and knowledge of perpendicular line slopes

Key Explanation: Choice D is correct. In order to find a line perpendicular to the given line, the slope of the original line must be determined first which yields

$m_{original} = \frac{\left(2-\frac{1}{2}\right)}{(4-2)} = \frac{\frac{3}{2}}{2} = \frac{3}{4}$. The slope of the perpendicular line is the negative reciprocal of the slope of the original line. Therefore, the slope of the perpendicular line is $-\frac{4}{3}$.

Next, to find the equation of the line in slope-intercept form ($y = mx + b$), find the value of y-intercept (b). To find (b), substitute the given point (3, –2) and the slope $\left(-\frac{4}{3}\right)$ to the slope-intercept form equation which yields $-2 = \left(-\frac{4}{3}\right)(3) + b$. Simplifying the equation yields $b = 2$. Therefore, the equation of the perpendicular line is $y = -\frac{4}{3}x + 2$.

Distractor Explanations: Choice A is incorrect and may result from an error in calculating the y-intercept value. **Choice B** is incorrect and may result from using the slope of the original line. **Choice C** is incorrect and may result from solving the equation of the original line.

24. **Level:** Hard | **Skill/Knowledge:** Linear equations in two variables | **Testing Point:** Converting English to algebraic equation with percentages

Key Explanation: Choice B is correct. To determine the equation of the total cost of purchase, begin by determining the total cost of (l) lamps and (r) rugs before the discount is applied which is $T_{before\ discount} = 75(r + l)$. Using distributive property, the equation then becomes $T_{before\ discount} = 75r + 75l$. When the 20% discount is applied, the cost is 80% of the original. This yields $T = 0.8(75r + 75l)$.

Distractor Explanations: Choice A is incorrect and is most likely the result of errors in applying discounts to equations with variables. **Choice C** is incorrect and is most likely the result of errors in applying discounts to equations with variables. **Choice D** is incorrect and is most likely the result of errors in applying discounts to equations with variables.

25. **Level:** Easy | **Skill/Knowledge:** Linear equations in two variables | **Testing Point:** Solving equation in two variables

Key Explanation: Choice B is correct. To solve for y, combine like terms on both sides of the equation as follows:

$$5y + 7y + 15y + 42y = 3y + 14y + z$$
$$69y = 17y + z.$$

Next, subtract $17y$ from both sides of the equation to get:
$52y = z$.

Then, divide both sides of the equation by 52 to get y as follows:

$y = \frac{z}{52}$. This gives **Choice B** as the correct answer.

Distractor Explanations: Choice A is incorrect and is likely the result of arithmetic errors. **Choice C** is incorrect and is likely the result of arithmetic errors. **Choice D** is incorrect and is likely the result of arithmetic errors.

26. **Level:** Easy | **Skill/Knowledge:** Linear equations in two variables | **Testing Point:** Solving linear equation given value of one variable

Key Explanation: Choice C is correct.
Substituting 2 for b yields $\frac{a-1}{2} = 2$.

Multiplying 2 to both sides of the equation yields $a - 1 = 4$.

Adding 2 to both sides of the equation yields $a + 1 = 6$.

Distractor Explanations: Choice A is incorrect and may result from solving the value of $a - 1$. **Choice B** is incorrect and may result from solving the value of a. **Choice D** is incorrect and may result from solving the value of $a + 2$.

27. **Level:** Hard | **Skill/Knowledge:** Linear equations in two variables | **Testing Point:** Solving a linear equation for one variable and evaluating an expression

Key Explanation: Choice B is correct.

Solving for n in the first equation yields $n = 6m - 24$.

Substituting the value of n in the expression $\frac{64^{2m}}{4^n}$ yields $\frac{64^{2m}}{4^{6m-24}}$. Changing the base in the numerator to 4 yields $\frac{\left(4^3\right)^{2m}}{4^{6m-24}} = \frac{4^{6m}}{4^{6m-24}}$. Since the terms have the same base and are being divided, the exponents will be subtracted. This yields $4^{6m-6m+24}=4^{24}$.

Distractor Explanations: Choice A is incorrect and reflects errors in using exponent rules to simplify equations. **Choice C** is incorrect and reflects errors in using exponent rules to simplify equations. **Choice D** is incorrect and reflects errors in using exponent rules to simplify equations.

28. **Level:** Easy | **Skill/Knowledge:** Linear equations in two variables | **Testing Point:** Evaluating a linear equation given one variable value

Key Explanation: Choice B is correct. Substitute 69 into the equation $y = 22 + 3.9x$ which yields $69 = 22 + 3.9x$. Solve for x by isolating x. Subtracting 22 from both sides of the equation yields $47 = 3.9x$. Dividing both sides of the equation by 3.9 yields $\frac{47}{3.9}=x$ or $x = 12.05$.

Distractor Explanations: Choice A is incorrect and may result from a conceptual or calculation error. **Choice C** is incorrect and may result from a conceptual or calculation error. **Choice D** is incorrect and may result from a conceptual or calculation error.

29. **Level:** Easy | **Skill/Knowledge:** Linear equations in two variables | **Testing Point:** Solving for one variable in terms of another in a linear equation

Key Explanation: Choice D is correct.

Since $x=\frac{7y-4}{4}$, solve for y by isolating the variable.

Multiplying both sides of the equation by 4 yields $4x = 7y - 4$.

Adding 4 to both sides of the equation yields $4x + 4 = 7y$ or $7y = 4x + 4$.

Dividing both sides of the equation by 7 yields $y=\frac{4x+4}{7}$.

Distractor Explanations: Choice A is incorrect and reflects errors in manipulating algebraic functions. **Choice B** is incorrect and reflects errors in manipulating algebraic functions. **Choice C** is incorrect and reflects errors in manipulating algebraic functions.

30

Daniel owns a website that sells tickets to live events. His website charges an initial, one-time fee for purchasing tickets through his website. The equation $C = 29.99s + 14.99$, where s is the number of tickets to purchase, represents the total amount of money that first-time buyers must pay. What does the number 29.99 represent in the equation?

A) The price per ticket, in dollars.

B) The initial fee that Daniel charges for first-time buyers.

C) The total amount that Daniel will earn per transaction.

D) The number of tickets purchased in the transaction.

31

A manager is calculating the amount of overtime pay she owes her employees. The equation $y = 20x + 25z$ represents the value of an employee's weekly paycheck in dollars (y), where x represents the number of regular hours worked and z represents the number of overtime hours worked. What is the best interpretation for the value 25 in the equation?

A) Total number of hours worked

B) Number of overtime hours

C) Rate of pay per normal hour of work

D) Rate of pay per overtime hour of work

32

A car maker operates production plants that saw an increase in production in 2018 to 5.6 million units annually from 5 million units in 2013. If the production of the cars increases at a constant rate, which function, f, in millions of units, best models the production of cars x years after 2013?

A) $f(x) = -\frac{3}{20}x + 5$

B) $f(x) = -\frac{3}{25}x + 5$

C) $f(x) = \frac{3}{20}x + 5$

D) $f(x) = \frac{3}{25}x + 5$

33

$$p(n) = 2n - 2$$

$$q(n) = 2 - p(n)$$

The functions p and q are defined above. What is the value of $q(2)$?

A) −4

B) −2

C) 0

D) 2

34

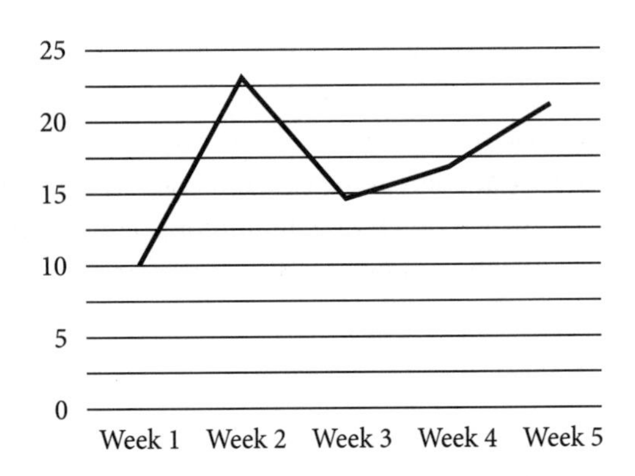

Dan collects baseball cards and tracks the number of baseball cards he has on the graph above. During which interval did the number of baseball cards increase the fastest?

A) Between Weeks 1–2

B) Between Weeks 2–3

C) Between Weeks 3–4

D) Between Weeks 4–5

35

An author sells the rights to her book to a publishing house. If she sold the book for an initial fee of $1,500 and she makes $18.50 for every purchase of the book (x), which expression represents her total profit from sales of the book?

A) $1{,}500 - 18.5x$

B) $18.5x - 1{,}500$

C) $18.5x + 1{,}500$

D) $18.5 + 1{,}500$

36

If $g(n) = 2n + 4$, what has the same value as $g(4) + g(6)$?

A) $g(8)$

B) $g(10)$

C) $g(12)$

D) $g(28)$

37

If $f(y) = \frac{6}{4}y + 6$ and $f(y) = 10$ what is the value of y^{-1}?

A) $\frac{3}{8}$

B) $\frac{5}{8}$

C) $\frac{8}{5}$

D) $\frac{8}{3}$

38

The function f is defined by $f(x) = 2x - 2$. Which of the following is equal to $f(x) + 2$?

A) $2x + 2$

B) $2x - 2$

C) $-2x$

D) $2x$

39

The cost of a bicycle rental on Beautiful Island is \$20, plus \$7 per hour for each hour that the bike is rented. If a bike is rented for x hours, which of the following functions of f represents the total cost in dollars to rent the bike?

A) $f(x) = 20 + 7x$

B) $f(x) = 7 - 20x$

C) $f(x) = 7x - 20$

D) $f(x) = -7x + 20$

40

If the equation of a line represented by the table below in the form $ax + by = c$, what is the value of b, where $b > 0$?

x	y
3	6
6	11
9	16

A) 1

B) $\frac{5}{3}$

C) 3

D) 5

41

Line p is perpendicular to line m. What is the slope of line p if line m passes through points (7, 3) and (2, 6)?

A) $-\frac{5}{3}$

B) $-\frac{3}{5}$

C) $\frac{3}{5}$

D) $\frac{5}{3}$

42

If $f(x) = -3x + 14$, what is the value of $f(4)$?

43

On a Sunday morning, Courtney sent x text messages each hour for 4 hours, and Lauren sent y text messages each hour for 6 hours. Which of the following represents the total number of messages sent by Courtney and Lauren on Sunday morning?

A) $10xy$

B) $24xy$

C) $6x + 4y$

D) $4x + 6y$

30. **Level:** Easy | **Skill/Knowledge:** Linear functions
Testing Point: Determining the meaning in context of coefficients of linear equations

Key Explanation: Choice A is correct. Since 29.99 is multiplied by the variable *s*, which represents the number of tickets to purchase, then it is the price per ticket in the equation.

Distractor Explanations: Choice B is incorrect. The initial fee is $14.99 because it is not multiplied by the number of tickets purchased. **Choice C** is incorrect because variable *C* represents the total amount that Daniel will earn per transaction. **Choice D** is incorrect because the variable *s* is the number of tickets purchased.

31. **Level:** Easy | **Skill/Knowledge:** Linear functions
Testing Point: Interpreting a value in a linear function

Key Explanation: Choice D is correct. In the equation, the variables *x* and *z* represent the number of hours worked by an employee. Since the total pay (*y*) is determined by multiplying *x* and *z* by 20 and 25 respectively, it can be deduced that 20 and 25 are the rates of pay. And since 25 is being multiplied by *z*, it is the rate of pay per overtime hour of work.

Distractor Explanations: Choice A is incorrect because $x + y$ represents the total number of hours worked. **Choice B** is incorrect because *z* represents the number of overtime hours. **Choice C** is incorrect because 20 represents the rate of pay per normal hour of work.

32. **Level:** Medium | **Skill/Knowledge:** Linear functions | **Testing Point:** Interpreting a linear function and finding the slope

Key Explanation: Choice D is correct. The function can be written in the slope-intercept form of an equation $f(x) = mx + b$, where m is the slope and b is the y-intercept. To calculate the slope use the slope formula m = the change in the y values divided by the change in the x values. There are 5 million units in 2013, at time equal to zero. This is thus the y-intercept and thus b. There was a 0.6 million unit increase over 5 years from 2013 to 2018. This indicates a slope $\frac{0.6}{5}=\frac{3}{25}$ *million units per 1 year*. Therefore, the function is $f(x)=\frac{3}{25}x+5$.

Distractor Explanations: Choice A is incorrect as it incorrectly shows a decrease in units produced each year. **Choice B** is incorrect as it incorrectly shows a decrease in units produced each year. **Choice C** is incorrect as it is the result of a math error involving calculating the slope.

33. **Level:** Easy | **Skill/Knowledge:** Linear functions
Testing Point: Evaluating linear functions

Key Explanation: The correct answer is **Choice C**. Substituting $p(n) = 2n - 2$ for $p(n)$ in the second equation and combining like terms yields $q(n) = 2 - p(n) = 2 - (2n - 2)$ or $2 - 2n + 2$ or $4 - 2n$.
Substituting 2 for n in $q(n) = 4n - 2$ results in $q(2) = 4 - 2(2) = 4 - 4 = 0$, which is **Choice C**.

Distractor Explanations: Choice A is incorrect and reflects error in knowledge of linear functions or simplification. **Choice B** is incorrect and reflects error in knowledge of linear functions or simplification. **Choice D** is incorrect and reflects error in knowledge of linear functions or simplification.

34. **Level:** Easy | **Skill/Knowledge:** Linear functions **Testing Point:** Interpreting a linear function from a graph

Key Explanation: Choice A is correct. Between Weeks 1 and 2, there was the steepest increase in slope on the graph.

Distractor Explanations: Choice B is incorrect and is the result of incorrectly interpreting the data in the graph or not reading the question carefully enough. **Choice C** is incorrect and is the result of incorrectly interpreting the data in the graph or not reading the question carefully enough. **Choice D** is incorrect and is the result of incorrectly interpreting the data in the graph or not reading the question carefully enough.

35. **Level:** Easy | **Skill/Knowledge:** Linear functions **Testing Point:** Interpreting terms of a linear function

Key Explanation: Choice C is correct. Since the author sold the book for $1,500, she has an initial profit of $1,500. Thereafter, each book sale adds $18.50 to her profits. This makes her total profit to be 1,500 + 18.5*x* or 18.5*x* + 1,500, where *x* is the number of books sold.

Distractor Explanations: Choice A is incorrect and may result if she loses $18.50 per book sale. **Choice B** is incorrect and may result if she paid $1,500 initially, rather than gaining $1,500. **Choice D** is incorrect and may result if she only sold one book, as there is no variable *x* in the expression.

36. **Level:** Medium | **Skill/Knowledge:** Linear functions | **Testing Point:** Evaluating a linear function given an input value

Key Explanation: Choice C is correct. $g(4)$ means to substitute 4 for n in the equation $g(n) = 2n + 4$, and likewise for $g(6)$. Therefore, the value of $g(4) + g(6) = [2(4) + 4] + [2(6) + 4]$. Simplifying the equation yields $g(4) + g(6) = [8 + 4] + [12 + 4] = 12 + 16 = 28$.

The question asks which has the same value as 28. To determine the answer, set up the equation equal to 28 and solve for n which yields $2n + 4 = 28$.

Subtracting 4 from both sides of the equation yields $2n = 24$.

Dividing both sides of the equation by 2 yields $n = 12$.

Therefore, $g(12) = g(4) + g(6)$.

Distractor Explanations: Choice A is incorrect and may result from following the pattern 4, 6, 8. **Choice B** is incorrect and may result from adding 4 and 6 together. **Choice D** is incorrect and may result from finding the value of $g(4) + g(6)$ but failing to find the value of n.

37. **Level:** Easy | **Skill/Knowledge:** Linear functions **Testing Point:** Evaluating a linear function and using negative exponent rule

Key Explanation: Choice A is correct. First, use the information provided to determine the value of y, before finding the value of $(y)^{-1}$.

Equating the values of $f(y)$ yields $\frac{6}{4}y+6=10$.

Subtracting 6 from both sides of the equation yields $\frac{6}{4}y=4$.

Multiplying 4 to both sides of the equation yields $6y = 16$.

Dividing both sides of the equation by 6 yields $y=\frac{16}{6}=\frac{8}{3}$.

Therefore, $y^{-1} = \frac{1}{y} = \frac{3}{8}$.

Distractor Explanations: Choice B is incorrect and may result from conceptual or calculation error. **Choice C** is incorrect and may result from a conceptual or calculation error. **Choice D** is incorrect and may result from solving the value of y.

38. **Level:** Easy | **Skill/Knowledge:** Linear functions
Testing Point: Evaluating a linear function

Key Explanation: Choice D is correct. Since $f(x) = 2x - 2$, then the expression $f(x) + 2 = 2x - 2 + 2 = 2x$.

Distractor Explanations: Choice A is incorrect and may result from solving $f(x) + 4$. **Choice B** is incorrect and may result from not adding 2 to the equation. **Choice C** is incorrect and may result from solving $-f(x) - 2$.

39. **Level:** Easy | **Skill/Knowledge:** Linear functions
Testing Point: Creating a linear function from information in question

Key Explanation: Choice A is correct. Start with the initial \$20 to rent the bike and add the additional fee per hour. Since the rate is \$7 per hour, the additional fee for renting a bike for x hours is expressed as $\$7 \times x = 7x$. Therefore, the function is defined by $f(x) = 20 + 7x$.

Distractor Explanations: Choice B is incorrect and is the result of errors in creating linear equations. **Choice C** is incorrect and is the result of errors in creating linear equations. **Choice D** is incorrect and is the result of errors in creating linear equations.

40. **Level:** Easy | **Skill/Knowledge:** Linear functions
Testing Point: Finding the equation of a line from data

Key Explanation: Choice C is correct. To find the equation of the line, we are required to find the slope of the line. Using (3, 6) and (6, 11) we can find it by $\frac{11-6}{6-3} = \frac{5}{3}$.

Substituting the slope to the equation of the line in slope-intercept form yields $y = \frac{5}{3}x + c$.

Plugging in point (3, 6) to the equation of the line yields $6 = \frac{5}{3}(3) + c$. Simplifying the equation yields $6 = 5 + c$. Subtracting 5 from both sides of the equation yields $1 = c$ or $c = 1$. Hence, the equation of the line is $y = \frac{5}{3}x + 1$. Rewrite the equation of the line in general form by multiplying 3 which yields $3\left(y = \frac{5}{3}x + 1\right)$ or $3y = 5x + 3$. Subtracting $5x$ from both sides of the equation yields $3y - 5x = 3$. Since b is the coefficient of y, $b = 3$.

Distractor Explanations: Choice A is incorrect. This is the value of the y-intercept of the line. **Choice B** is incorrect. This is the slope of the line. **Choice D** is incorrect. This is the negative value of a.

41. **Level:** Easy |**Skill/Knowledge:** Linear functions
Testing Point: Finding the perpendicular slope of a line given two points

Key Explanation: Choice D is correct. The slope of line m given can be found by $\frac{y_2 - y_1}{x_2 - x_1}$.

Substituting the given points yields $\frac{6-3}{2-7}$ or

$-\frac{3}{5}$. A slope of line perpendicular to this can be found by the negative inverse of this slope. The slope of line p would therefore be $\frac{5}{3}$.

Distractor Explanations: Choice A is incorrect and may result from not getting the negative value of the slope of line m. **Choice B** is incorrect. This is the slope of a line m or a line parallel to m. **Choice C** is incorrect and may result from not getting the inverse of the slope of line m.

42. **Level:** Easy | **Skill/Knowledge:** Linear functions
Testing Point: Using the value of x to find $f(x)$

Key Explanation: Substitute x for 4 in the given equation.

This yields $f(4) = -3(4) + 14$. Simplifying the equation yields $f(4) = 2$.

43. **Level:** Easy | **Skill/Knowledge:** Linear functions
Testing Point: Creating a linear expression from data given

Key Explanation: Choice D is correct. The number of text messages Courtney sent on Sunday morning was $4x$. The number of text messages Lauren sent on Sunday morning was $6y$. To find the total amount, find the sum of Courtney's and Lauren's number of text which is $4x + 6y$.

Distractor Explanations: Choice A is incorrect and may result from a conceptual error. **Choice B** is incorrect and may result from getting the product of Courtney's and Lauren's number of texts. **Choice C** is incorrect and may result from swapping the respective variables for Courtney's and Lauren's number of texts.

44

$$-x + 2y = -8$$
$$4x - y = 25$$

What is the solution (x, y) to the system of equations above?

A) $(-1, 6)$

B) $(6, -1)$

C) $(1, -6)$

D) $(6, -6)$

45

A farmer's market is selling spots for their next event. They have 64 total square feet available for rent, with two options in size: 6 square feet and 4 square feet. If the market rented out a total of 14 spots, which system of equations below represents the relationship between the number of 6 square feet spots sold (s) and 4 square feet spots sold (f)?

A) $6s + 4f = 14,\ s + f = 64$

B) $6f + 4s = 64,\ s - f = 14$

C) $6s + 4f = 64,\ s + f = 14$

D) $4s + 6f = 14,\ sf = 64$

46

$$6a - 4b = 24$$
$$-a + 2b = -8$$

If (a, b) is the solution to the system of equations above, what is the value of $a + b$?

A) -6

B) -1

C) $-\frac{2}{3}$

D) 5

47

$$3x + y = 9$$
$$-x + 5y = 21$$

In the system of equations above, what is the value of $x + y$?

48

$$-x + 2y = 8.1$$
$$5x - 3y = -\frac{122}{5}$$

The system of equations above can be solved by solution (x, y). What is the value of $2x - y$?

A) -9.3

B) -3.5

C) 2.3

D) 7

Systems of two linear equations in two variables

49

A group of 100 students went on a trip for a choir competition and lodged in 20 hotel suites that either held 4 or 6 students. If all of the rooms were filled to capacity, what is the difference between the number of 4 and 6 person suites?

50

$$2x - 4y = -17$$
$$-x + 3y = \frac{27}{2}$$

What is the solution to the system of equations above?

A) (−5, 1.5)

B) (15, 5)

C) (3, −3)

D) (1.5, 5)

51

$$-3x + 4y = s$$
$$rx - 8y = -28$$

What values of (r, s) make the system of equations above have infinitely many solutions?

A) (6, 14)

B) (14, 6)

C) (3, 4)

D) (9, 5)

52

In the xy-plane, the lines $4x - 2y = 6$ and $y = mx + b$ are perpendicular. What is the value of m?

53

In the xy-plane, the line $-9x + 3y = 81$ intersects the y-axis at $y = b$. What is the value of b?

A) −9

B) 3

C) 9

D) 27

54

$$2x - y = -8$$
$$-x + 4y = 25$$

What is the solution (x, y) to the system of equations above?

A) (−1, 6)

B) (6, −1)

C) (1, −6)

D) (6, −6)

55

If (x, y) are solutions to the system below, what is the value of $x - y$?

$$y = -\frac{2}{3}x$$

$$3x - 2y = -13$$

A) −6

B) −5

C) −3

D) 2

56

$$6x - 10y = 5$$

If the equation above is one of the equations in a linear system with infinite solutions, which of the following equations is a part of that system?

A) $-12x + 20y = -10$

B) $12x + 10y = 10$

C) $12x - 20y = 20$

D) $12x + 20y = -10$

57

A test has 22 questions in which there are m Multiple Choice Questions and s Student Produced Responses. The Multiple Choice Questions are 3 points each and Student Produced Response questions are 2 points each. If the maximum score on a test is 61 points, how many MCQs are on the test?

A) 5

B) 10

C) 17

D) 51

58

$$24x - 6y = 33$$

$$8x - sy = -12$$

For what values of s does the system of equations above have no real solutions?

59

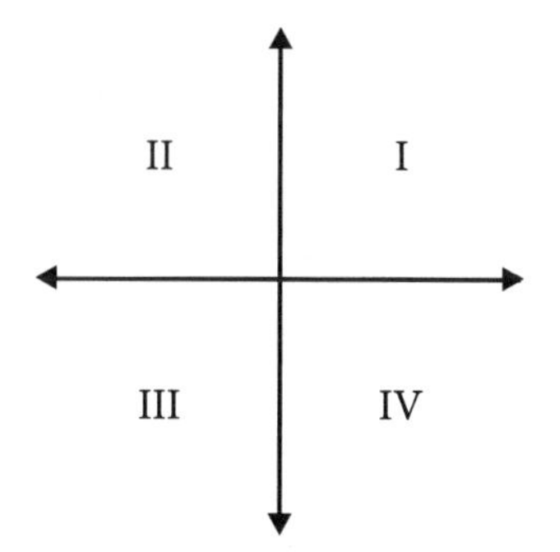

$$9x + y = -19$$

$$\frac{7}{4}x - 5y = \frac{3}{2}$$

For the system of equations above, in which quadrant does the solution lie?

A) I

B) II

C) III

D) IV

60

$$\frac{4}{5}x + by = 17$$

$$ax - 4y = c$$

If the system of equations above has infinitely many solutions, what is the value of $a + b + c$, if $b = 2$?

A) -34

B) -33.6

C) $-\frac{16}{5}$

D) $-\frac{8}{5}$

61

$$dx = -7y + 4$$

$$4x + 14y = e$$

In the system of equations above, for what value of d is the system true for all real numbers?

62

Jerry is buying cups and napkins for a party which costs \$3 and \$2 each, respectively. If he buys twice as many napkins as cups, how many cups did he buy if he spent \$105 in total? Note: disregard tax in your answer.

63

Jennifer is putting together panels of speakers for a conference. There are two panels, one that is 60 minutes and another one that is 120 minutes. Jennifer is offering speakers on the 60-minute panel \$50 for their time, and the speakers on the 120-minute panel \$110 for their time. If Jennifer budgets \$900 for 12 total speakers, which of the following systems of equations represents the number of 60-minute panelists (x) and 90-minute panelists (y)?

A) $x + y = 12$, $50x + 110y = 900$

B) $x + y = 900$, $50x + 110y = 12$

C) $xy = 12$, $50y + 110x = 900$

D) $x - y = 900$, $50x + 110y = 12$

64

Is (25, 11) the solution to the following system of equations?

$$y + x = 36$$
$$2y + 8x = 140$$

A) Yes, (25, 11) is the solution because it satisfies both equations.

B) No, (25, 11) is not the solution because it does not satisfy the first equation.

C) No, (25, 11) is not the solution because it does not satisfy the second equation.

D) More information is needed in order to answer the question.

65

$$4e + f = 2$$
$$e - f = 3$$

If the point (e, f) satisfies the system of equations above, what is the value of e?

66

Quentin's sales: $T_Q = 5b + 48$

Alice's sales: $T_A = 5.5b + 159$

Quentin and Alice sell bagels before school. The equations above represent the number of bagels (b) they sell on a given day. How many bagels did they sell on a given day if Alice made twice as much money as Quentin?

67

According to the preceding system of equations, what is the value of y? (round the answer to two decimal places)

$$y - 2x = -6$$
$$2y - x = 5$$

68

If the system of equations below has no solution, what is the value of p?

$$y = 3(x - 5) - 2(-5 + x)$$
$$y = p(2 - x) + 6x$$

44. **Level:** Easy | **Skill/Knowledge:** Systems of two linear equations in two variables | **Testing Point:** Finding the correct value by elimination

 Key Explanation: Choice B is correct. Use an elimination method to solve for the value of x and y. To eliminate y, first, multiply the bottom equation by 2 which yields $8x - 2y = 50$. Adding the two equations yields $7x = 42$. Dividing both sides of the equation by 7 yields $x = 6$. By plugging $x = 6$ into the first equation yields $-(6) + 2y = -8$. Adding 6 and dividing both sides of the equation by 2 yields $y = \frac{-8+6}{2}$ or $y = -1$. Therefore, the solution to the systems of equation is (6, −1).

 Distractor Explanations: Choice A is incorrect and may result from interchanging the values of x and y. **Choice C** is incorrect and may result from interchanging the values and signs of x and y. **Choice D** is incorrect and may result from a conceptual or calculation error.

45. **Level:** Easy | **Skill/Knowledge:** Systems of two linear equations in two variables | **Testing Point:** Creating a system of two linear equations from data in the problem

 Key Explanation: Choice C is correct. In order to create equations that represent the data from the market, the variables s and f must be matched and multiplied with their correct totals and constants. 6 square feet is denoted by s and hence total 6 square feet rented area is 6 square feet × s = $6s$. 4 square feet is denoted by f and hence total 4 square feet rented area is 4 square feet × $f = 4f$. If the total number of spots sold is 14, then the first equation will be $s + f = 14$. If the total rented area is 64, then the second equation will be $6s + 4f = 64$.

 Distractor Explanations: Choice A is incorrect because it has interchanged the total area rented and total spots rented. **Choice B** is incorrect as instead of adding the number of rented spots, it is subtracting them. **Choice D** is incorrect because it multiplies the area with wrong variables.

46. **Level:** Easy | **Skill/Knowledge:** Systems of two linear equations in two variables | **Testing Point:** Solving a system of two linear equations

 Key Explanation: Choice B is correct. The system of equations can be solved by either substitution or elimination, but the standard form of the equation indicates that elimination is best. Multiplying 2 to both sides of the second equation yields $-2a + 4b = -16$. Adding the two equations yields $6a - 4b - 2a + 4b = 24 - 16$. Combining like terms yields $4a = 8$. Dividing both sides of the equation by 4 yields $a = 2$. Substituting the value of a to the original second equation yields $-(2) + 2b = -8$. Adding 2 to both sides of the equation yields $2b = -6$. Dividing both sides of the equation by 2 yields $b = -3$. Therefore, the solution to the system of equations is (2, −3). Adding the values of a and b yields $2 - 3 = -1$.

 Distractor Explanations: Choice A is incorrect and may result from calculating the product of a and b. **Choice C** is incorrect and may result from calculating the quotient of a and b. **Choice D** is incorrect and may result from calculating the difference between a and b.

47. **Level:** Medium | **Skill/Knowledge:** Systems of two linear equations in two variables | **Testing Point:** Creating and using a system of two linear equations

 Key Explanation: Use the elimination method to solve for x and y. Multiplying 3 to both sides of the second equation yields $3(-x + 5y) = 21(3)$. Using distributive property yields $-3x + 15y = 63$.

Adding the first and the second equation yields $3x + y - 3x + 15y = 9 + 63$. Combining like terms yields $16y = 72$. Dividing both sides of the equation by 16 yields $y = \frac{72}{16} = 4.5$.
To determine x, substitute 4.5 for y in the first equation which yields $3x + 4.5 = 9$. Subtracting 4.5 from both sides of the equation yields $3x = 4.5$. Dividing 4 from both sides of the equation yields $x = 1.5$.

Therefore, the sum of x and y is $x + y = 1.5 + 4.5 = 6$.

48. **Level:** Medium | **Skill/Knowledge:** Systems of two linear equations in two variables | **Testing Point:** Solving a system of two linear equations

Key Explanation: Choice A is correct. Start by solving the system of equations using the elimination method. Multiplying the first equation by 5 yields $-5x + 10y = 40.5$. Adding the first and the second equation yields $-5x + 10y + 5x - 3y = 40.5 - \frac{122}{5}$. Combining like terms yields $7y = 16.1$. Dividing 7 from both sides of the equation yields $y = 2.3$.

Then, substitute this value into the original first equation and solve for x. This yields $-x + 2(2.3) = 8.1$ or $x = -3.5$.

Therefore, $2x - y = 2(-3.5) - 2.3 = -9.3$.

Distractor Explanations: Choice B is incorrect and most likely result from arithmetic errors in solving the system of equations. **Choice C** is incorrect and most likely result from arithmetic errors in solving the system of equations. **Choice D** is incorrect and most likely result from arithmetic errors in solving the system of equations.

49. **Level:** Medium | **Skill/Knowledge:** Systems of two linear equations in two variables | **Testing Point:** Creating and using a system of two linear equations

Key Explanation: 0 is correct. Let x represent the number of 4 person suites and let y represent the number of 6 person suites. Since the number of hotel suites is 20, $x + y = 20$. Since there are 4 people in each x suite and 6 people in each y suite, $4x + 6y = 100$. Solving for x in the first equation $x + y = 20$, results in $x = 20 - y$. Substituting this value for x into the second equation results in $4(20 - y) + 6y = 100$. Multiplying the 4 by $(20 - y)$ results in $80 - 4y + 6y = 100$. Combining like terms gets $80 + 2y = 100$. Subtracting 80 from both sides of the equation results in $2y = 20$. Therefore $y = 10$. Since $y = 10$, solving for x by substituting the value for y into the first equation $x + y = 20$, results in $x + 10 = 20$, or $x = 10$.

The difference between the x and y values is $10 - 10 = 0$.

50. **Level:** Medium | **Skill/Knowledge:** Systems of two linear equations in two variables | **Testing Point:** Solving a system of two linear equations

Key Explanation: Choice D is correct.

Eliminate one of the variables. Multiplying 2 to the 2nd equation yields $2(-x + 3y) = 2\left(\frac{27}{2}\right)$ or $-2x + 6y = 27$.

Adding the two equations yields $2x - 4y - 2x + 6y = -17 + 27$.

Combining like terms yields $2y = 10$.

Dividing both sides of the equation by 2 yields $y = 5$.

Substitute this value for y in one of the original equations and solve for x.

Substituting 5 to the 1st equation yields $2x - 4(5) = -17$ or $2x - 20 = -17$.

Adding 20 to both sides of the equation yields $2x = 3$.

Dividing both sides of the equation by 2 yields
$x = \frac{3}{2} = 1.5.$
Therefore, the solution to the system of equations is (1.5, 5).

Distractor Explanations: Choice A is incorrect and may result from a conceptual or calculation error. **Choice B** is incorrect and may result from a conceptual or calculation error. **Choice C** is incorrect and may result from a conceptual or calculation error.

51. **Level:** Medium | **Skill/Knowledge:** Systems of two linear equations in two variables | **Testing Point:** Solving a system of two linear equations with infinite solutions

    Key Explanation: Choice A is correct. Systems of equations that have infinitely many solutions are essentially the same line, so the constants and the coefficients must be proportional to one another. Using the coefficients of the y-values, the proportion is 1: −2. Therefore, the coefficients of the x and the constants have also the same ratio.

    Multiplying the coefficient of x in the 1st equation to get r yields $r = -3(-2)$ or $r = 6$.

    Dividing the constant in the 2nd equation to get s yields $s = \frac{-28}{-2}$ or $s = 14$.

    The solution is therefore (6, 14).

    Distractor Explanations: Choice B is incorrect and may result from swapping the values of s and r. **Choice C** is incorrect and may result from a conceptual or calculation error. **Choice D** is incorrect and may result from a conceptual or calculation error.

52. **Level:** Medium | **Skill/Knowledge:** Systems of two linear equations in two variables | **Testing Point:** Using a system of two linear equations with perpendicular lines

    Key Explanation: To find the slope of $4x - 2y = 6$, express the linear equation in slope-intercept form. Subtracting $4x$ from both sides of the equation yields $-2y = -4x + 6$. Dividing both sides of the equation by −2 yields $y = 2x - 3$. The slope is 2, so a perpendicular line would have a negative reciprocal slope. Change the sign from negative to positive and flip the fraction. It will yield $m = -\frac{1}{2}$ or −0.5.

53. **Level:** Easy | **Skill/Knowledge:** Systems of two linear equations in two variables | **Testing Point:** Solving a system of two linear equations

    Key Explanation: Choice D is correct. Rewrite the linear equation $-9x + 3y = 81$ in slope-intercept form. Adding $9x$ on both sides of the equation yields $3y = 9x + 81$. Dividing both sides of the equation by 3 yields $y = 3x + 27$. Since the slope-intercept form is $y = mx + b$ and b is the y-intercept, $b = 27$.

    Distractor Explanations: Choice A is incorrect and reflects errors in interpreting linear equations. **Choice B** is incorrect and reflects errors in interpreting linear equations. **Choice C** is incorrect and reflects errors in interpreting linear equations.

54. **Level:** Easy | **Skill/Knowledge:** Systems of two linear equations in two variables | **Testing Point:** Solving a system of linear equations by elimination or substitution

    Key Explanation: Choice A is correct. To eliminate one variable, first, multiply the 2nd equation by 2 so that the coefficients of x have opposite signs which yield $-2x + 8y = 50$. Adding the two equations together yields $2x - y - 2x +$

$8y = -8 + 50$ or $7y = 42$. Dividing both sides of the equation by 7 yields $y = 6$. Plugging $y = 6$ into the 1st equation yields $2x - (6) = -8$. Adding 6 to both sides of the equation yields $2x = -2$. Dividing both sides of the equation by 2 yields $x = -1$. Therefore, the solution to the system of equations is $(-1, 6)$.

Distractor Explanations: Choice B is incorrect and may result from swapping the values of x and y. **Choice C** is incorrect and may result from swapping the signs of x and y. **Choice D** is incorrect and may result from a conceptual or calculation error.

55. **Level:** Medium | **Skill/Knowledge:** Systems of two linear equations in two variables | **Testing Point:** Solving for linear systems simultaneously

Key Explanation: Choice B is correct. Solve the system of equations using substitution.

Substituting the 1st equation to the 2nd equation yields $3x - 2\left(-\frac{2}{3}x\right) = -13$. Using the distributive law and grouping like terms to solve for x would yield $\frac{13}{3}x = -13$. Multiplying 3 and dividing both sides of the equation by 13 yields $x = -3$. Substituting the value of x in the 1st equation to find the value of y yields $y = -\frac{2}{3}(-3)$ or $y = 2$.

Substituting the values of x and y to the expression $x - y$ yields $-3-2 = -5$.

Distractor Explanations: Choice A is incorrect and is the product of x and y. **Choice C** is incorrect and is the value of x. **Choice D** is incorrect and is the value of y.

56. **Level:** Hard | **Skill/Knowledge:** Systems of two linear equations in two variables | **Testing Point:** Solving linear systems with infinite solutions

Key Explanation: Choice A is correct. For a system to have infinite solutions, the linear equations must be the same line written differently. This means that the slope and y-intercepts of both lines are same. Multiplying the given equation by -2 yields $-12x + 20y = -10$ which is **Choice A**.

Distractor Explanations: Choice B is incorrect. This system would have one solution as the linear equations would have different slopes. **Choice C** is incorrect. This system would have no solution as the lines would be parallel. **Choice D** is incorrect. This system would have one solution as the linear equations would have different slopes.

57. **Level:** Easy | **Skill/Knowledge:** Systems of two linear equations in two variables | **Testing Point:** Creating and solving for a system of linear equations

Key Explanation: Choice C is correct. There are a total of 22 questions

$$m + s = 22$$
$$3m + 2s = 61.$$

Solve for the system of equations simultaneously. Multiply the 1st equation by 2, which would result in $2m + 2s = 44$. Solve by subtracting the two equations which yields $m = 17$.

Distractor Explanations: Choice A is incorrect. This is the number of the Student Produced Response questions. **Choice B** is incorrect. This is the value of the total points that could be gained from Student Produced Responses. **Choice D** is incorrect. This is the value of the total points that could be gained from Multiple Choice Questions.

58. **Level:** Hard | **Skill/Knowledge:** Systems of two linear equations in two variables | **Testing Point:** Comparing linear equations in a system with no

real solutions

Key Explanation: The correct answer is 2. A system of equations with no solution on a coordinate plane is two parallel lines or lines with the same slope. To solve for the slope, convert each equation to slope-intercept form. Subtracting $24x$ to both sides of the first equation yields $-6y = -24x + 33$. Dividing -6 from both sides of the first equation yields $y = \frac{-24}{-6}x + \frac{33}{-6}$ or $y = 4x - \frac{33}{6}$. Therefore, the slope of the first equation is 4. Subtracting $8x$ to both sides of the second equation yields $-sy = -8x - 12$. Dividing $-s$ from both sides of the second equation yields $y = \frac{-8}{-s}x - \frac{12}{-s}$ or $y = \frac{8}{s}x + \frac{12}{s}$. Therefore, the slope of the second equation is $\frac{8}{s}$. Since the two equations have equal slope, then $4 = \frac{8}{s}$. Multiplying s and dividing 4 from both sides of the equation yields $s = \frac{8}{4}$. Therefore, $s = 2$.

59. **Level:** Medium | **Skill/Knowledge:** Systems of two linear equations in two variables | **Testing Point:** Solving a system of linear equations and determining the quadrant of the solution

Key Explanation: Choice C is correct. First, find the solution to the system of equations using the elimination method. Multiplying the first equation by 5 yields $45x + 5y = -95$. Adding the first and the second equation yields $45x + 5y + \frac{7}{4}x - 5y = -95 + \frac{3}{2}$. Combining like terms yields $\frac{187}{4}x = -\frac{187}{2}$. Multiplying 4 and dividing 187 from both sides of the equation yields $x = -2$. Substituting x to the original first equation yields $9(-2) + y = -19$. Simplifying the equation yields $y = -1$. Since both x and y values are negative, the solution lies on the third quadrant.

Distractor Explanations: Choice A is incorrect and most likely results from calculation errors in solving the system of equations. **Choice B** is incorrect and most likely results from calculation errors in solving the system of equations. **Choice D** is incorrect and most likely results from calculation errors in solving the system of equations.

60. **Level:** Easy | **Skill/Knowledge:** Systems of two linear equations in two variables | **Testing Point:** Solving a system of equations with infinitely many solutions

Key Explanation: Choice B is correct. For a system to have infinitely many solutions, the coefficients of both x and y and the constant on the right side of the equation must be proportional between the two equations. Using the value of b and the coefficient of y in the second equation to calculate the ratio between the first and second equation yields $\frac{-4}{2}$ or -2. Therefore, the coefficients and constant in the first equation must be multiplied by -2 to get the coefficients and constant in the second equation. Multiplying -2 to get the value of a yields $\frac{4(-2)}{5}$. Therefore, $a = -\frac{8}{5}$ or -1.6. Multiplying -2 to get the value of c yields $17(-2)$. Therefore, $c = -34$. Adding the values of a, b, and c yields $a + b + c = -1.6 + 2 - 34 = -33.6$.

Distractor Explanations: Choice A is incorrect because it is the value of c. **Choice C** is incorrect and may result from a conceptual or calculation error. **Choice D** is incorrect because it is the value of a.

61. **Level:** Medium | **Skill/Knowledge:** Systems of two linear equations in two variables | **Testing**

Point: Solving systems of linear equations in two variables

Key Explanation: 2 is correct. In order for a system of equations to be true for all real numbers, the equations must be the same line. The coefficients for both variables, as well as the constants, must be proportional to one another. Start by reorganizing the equations into the same form which yields

$dx + 7y = 4$ and

$4x + 14y = e$.

The coefficients of y reveal that the equations are proportional and have a ratio of 7:14 or 1:2. Therefore, the coefficients of x must also have the same ratio. Applying the ratio yields $\frac{d}{4} = \frac{1}{2}$. Multiplying 4 to both sides of the equation yields $d = 2$.

62. **Level:** Easy | **Skill/Knowledge:** Systems of two linear equations in two variables | **Testing Point:** Creating and solving a system of linear equations in two variables

Key Explanation: The correct answer is 15 cups. To determine the number of cups Jerry bought, one must write a system of equations, with variables for the number of napkins bought (n) and cups bought (c).

Since he buys twice as many napkins as cups, the first equation can be written as $2c = n$.

Since he spent a total of $105 for buying napkins and cups, the second equation can be written as $2n + 3c = 105$.

Substituting the value of n from the first equation to the second equation yields $2(2c) + 3c = 105$ or $4c + 3c = 105$.

Combining like terms yields $7c = 105$.

Dividing both sides of the equation by 7 yields $c = 15$.

63. **Level:** Easy | **Skill/Knowledge:** Systems of two linear equations in two variables | **Testing Point:** Creating a system of equations from text in a word problem

Key Explanation: Choice A is correct. The word problem describes a system of equations with variables representing the number of 60-minute panelists x and 90-minute panelists y. If Jennifer is offering speakers on the 60-minute panel $50 for their time, and the speakers on the 120-minute panel $110 for their time, then one equation must be $50x + 110y = 900$. The other equation represents the total number of speakers $x + y$, which must be equal to 12. Therefore, the other equation is $x + y = 12$.

Distractor Explanations: Choice B is incorrect and interchanges the value of total budget and total number of speakers. **Choice C** is incorrect as it multiples the number of 60-minute speakers and number of 120-minute speakers. **Choice D** is incorrect as it interchanges total number of speakers and budget, and also subtracts the number of 60-minute speakers and 120-minute speakers.

64. **Level:** Easy | **Skill/Knowledge:** Systems of two linear equations in two variables | **Testing Point:** Solving a system of linear equations

Key Explanation: Choice C is correct. Plugging (25, 11) into the first equation, yields $11 + 25 = 36$, which is true. Plugging (25, 11) into the second equation yields $2(11) + 8(25) = 140$. Simplifying the equation yields $222 = 140$, which is false. Therefore, **Choice C** is the correct answer.

Distractor Explanations: Choice A is incorrect because the solution does not satisfy second equation. **Choice B** is incorrect because the solution satisfies the first equation. **Choice D** is incorrect. There is enough information to answer the question.

65. **Level:** Easy | **Skill/Knowledge:** Systems of two linear equations in two variables | **Testing Point:** Solve a system of linear equations

    Key Explanation: The correct answer is 1. The easiest way to solve the system of equations is to use the elimination method. Adding the two equations directly yields $5e = 5$. Dividing both sides of the equation by 5 results in $e = 1$.

66. **Level:** Medium | **Skill/Knowledge:** Systems of two linear equations in two variables | **Testing Point:** Solving a system of linear equations

    Key Explanation: 14 is the correct answer. First multiply Quentin's sales by two in order to set the entire equation equal to Alice's sales:

    $2(5b + 48) = 5.5b + 159$.

    Using distributive property yields $10b + 96 = 5.5b + 159$.

    Subtracting 96 and $5.5b$ from both sides of the equation yields $4.5b = 63$.

    Dividing both sides of the equation by 4.5 yields $b = 14$.

67. **Level:** Easy | **Skill/Knowledge:** Systems of two linear equations in two variables | **Testing Point:** Solving a system of two linear equations

    Key Explanation: The correct answer is 5.33. Using the first equation to solve for y yields $y = -6 + 2x$. Substitute the value for y in terms of x into the second equation and solve for x. This yields $2(-6 + 2x) - x = 5$. Using distributive property yields $-12 + 4x - x = 5$. Combining like terms yields $-12 + 3x = 5$. Adding 12 to both sides of the equation yields $3x = 17$. Dividing both sides of the equation by 3 yields $x = \frac{17}{3}$. Now, find y by substituting the value for x into either of the two equations. Substituting it to the first equation yields $y = 2\left(\frac{17}{3}\right) - 6 = \frac{34}{3} - 6 = \frac{16}{3} = 5.33$.

68. **Level:** Medium | **Skill/Knowledge:** Systems of two linear equations in two variables | **Testing Point:** Working with linear systems with no solutions

    Key Explanation: 5 is the correct answer. Linear systems with no solution are parallel and have the same gradient.

    Using distributive property to simplify both equations yields $y = 3x - 15 + 10 - 2x$ and $y = 2p - px + 6x$.

    Combining like terms yields $y = x - 5$ and $y = (6 - p)x + 2p$. Equating the slopes of the equations yields $1 = 6 - p$. Adding p and subtracting 1 from both sides of the equation yields $p = 5$.

69

If $x \geq 3$, which of the following statements is NOT true?

A) $2x \geq 6$

B) $4x + 2 \geq 14$

C) $\frac{6}{5}x \geq 21$

D) $\frac{1}{2}x - 2 \geq -\frac{1}{2}$

70

The shaded region in the graph represents which of the following inequalities?

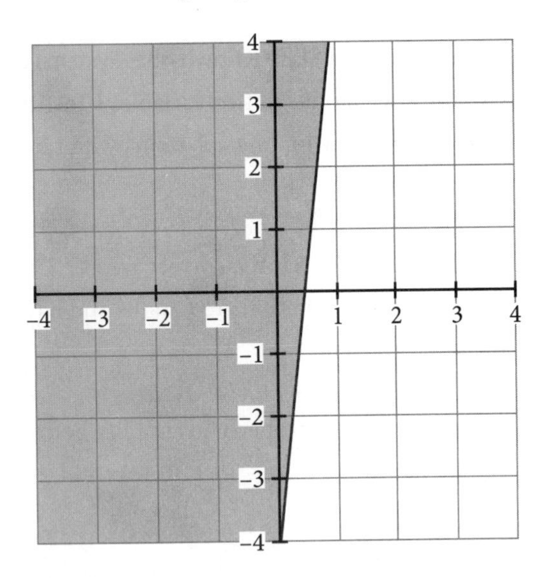

A) $y \geq 9x - 4$

B) $y > 9x - 4$

C) $y < 9x - 4$

D) $y \leq 9x - 4$

71

If $8y - 4x = 5$, and $y > 800$, then what is the least integer value of x?

72

Jackie is scheduling her next haircut. If her hair is already 10 inches long and she knows her hair grows at least an inch every two months, which statement represents the total length of her hair (y) after (x) months?

A) $y \geq 0.5x + 10$

B) $y = 0.5x + 10$

C) $y \geq 2x + 10$

D) $y \geq 0.5x$

73

If $x \geq 40$ and $y \leq 30$, then which of the following must also be true?

A) $x - y \geq 70$

B) $x - y \leq 10$

C) $x - y \geq 10$

D) $x + y \leq 10$

Linear inequalities in one or two variables

74

If $-40 \le x \le 90$, and $-30 \le y \le 20$, then what is the greatest possible value of $x - y$?

75

If $|6x + 14| \ge 4x + 24$, then which of the following is true?

A) $x \ge 5$

B) $x \le -\frac{19}{5}$

C) $-\frac{19}{5} \le x \le 5$

D) $x \ge 5$ or $x \le -\frac{19}{5}$

76

If $\sqrt{2{,}016} < a\sqrt{14}$, $\frac{a}{\sqrt{14}} < \sqrt{14}$, and a is an integer, then what is the value of a?

77

$$x + 2 \le -1$$

$$x > \frac{6}{3}x + 2$$

Which of the following inequalities expresses the domain of x-coordinates that satisfy the system of inequalities above?

A) $x \le -1$

B) $x \ge -1$

C) $-2 < x \le -3$

D) $x \le -3$

78

If $y \ge 4$, which of the following statements is NOT true?

A) $3y \ge 12$

B) $4y + 7 \ge 22$

C) $\frac{8}{5}y \ge 25$

D) $\frac{1}{2}y - 2 \ge -\frac{1}{2}$

79

Jane is knitting scarves and hats for her family as presents. If each scarf takes 15 feet of yarn to make and each hat takes 12 feet of yarn to make, and she purchases 425 feet of yarn, which inequality represents the maximum number of hats (h) and scarves (s) she can make for her family?

A) $425 \geq 15h + 12s$

B) $425 \geq 15s + 12h$

C) $425 \leq 15h + 12s$

D) $425 \leq 15s + 12h$

80

A hardware company is selling a new kind of picture hanging stud that can hold up to 12 pounds. If a painting is x pounds and the frame is y pounds, which of the following inequality correctly expresses the amount of weight that the stud(s) can hold?

A) $s \geq y - x$

B) $s \geq y + x$

C) $s \leq y - x$

D) $s \leq y + x$

81

What is the value of t if t is an integer and $|2t - 2| < 1$?

A) -1

B) 0

C) 1

D) 2

82

If $pq < 100$ and q is a positive multiple of 6, what is the greatest possible integer value of p?

A) 6

B) 16

C) 30

D) 96

83

Variables r and s are real numbers. If $20 < r < 30$ and $40 < s < 60$, which expression represents all the possible values of the expression $r - s$?

A) $-40 < r - s < -10$

B) $-30 < r - s < -20$

C) $-30 < r - s < 20$

D) $-20 < r - s < 30$

84

Which of the following is a solution to the inequality below?

$$2x + 5 \leq 3x + 7$$
$$2y > -x + 4$$

A) (2, 4)

B) (−4, 3)

C) (2, 1)

D) (−3, 3)

85

A bakery has a single oven for making loaves of bread and cakes. Using the oven, the baker can only make 27 pastries in a day. The baker can only bake a maximum of 4 loaves of bread in a day. If the bakery must produce at least one for each type of pastry, what is the maximum number of cakes the baker can make on a particular day?

A) 23

B) 24

C) 26

D) 27

86

If $12n < 49$ and $7n > 22$ and n is an integer, what is the value of n?

87

Carmen wants to spend no more than $43 on books and pens for her next semester. A box of pens costs $5 and a book costs $3. If she buys only one box of pens, what is the maximum number of books she can buy?

A) 3

B) 12

C) 13

D) 38

69. **Level:** Medium | **Skill/Knowledge:** Linear inequalities in one or two variables | **Testing Point:** Solving linear inequalities

Key Explanation: Choice C is correct. In order to find the inequality that is not true, simplify each option by isolating x. The inequality that will not be simplified as $x \geq 3$ will be the answer.

Simplify **Choice C** by multiplying 5 and dividing 6 from both sides of the inequality. This yields $x \geq \frac{5(21)}{6}$ or $x \geq \frac{35}{2}$. Therefore, this statement is false.

Distractor Explanations: Choice A is incorrect because the inequality can be simplified to $x \geq 3$ by dividing both sides by 2. **Choice B** is incorrect because the inequality can be simplified to $x \geq 3$ by subtracting 2 then dividing both sides by 4. **Choice D** is incorrect because the inequality can be simplified to $x \geq 3$ by adding 2 and then multiplying both sides by 2.

70. **Level:** Easy | **Skill/Knowledge:** Linear inequalities in one or two variables | **Testing Point:** Determining a linear inequality equation from its graph

Key Explanation: Choice A is correct. The solid line on the graph indicates that the inequality symbol used is either ≥ or ≤ sign. Since the shading on the graph is above the line, then it indicates that the inequality symbol used is either > or ≥. Since ≥ appears on both criteria, then this is the symbol used in the inequality represented by the graph.

Distractor Explanations: Choice B is incorrect and most likely results from a lack of knowledge in graphing inequalities. **Choice C** is incorrect and most likely results from a lack of knowledge in graphing inequalities. **Choice D** is incorrect and most likely results from a lack of knowledge in graphing inequalities.

71. **Level:** Easy | **Skill/Knowledge:** Linear inequalities in one or two variables | **Testing Point:** Solving linear inequalities

Key Explanation: 1,599 is the correct answer. Using the given equation $8y - 4x = 5$, find the value of y in terms of x.

Adding $4x$ to both sides of the equation yields $8y = 4x + 5$.

Dividing both sides of the equation by 8 yields $y = \frac{4x+5}{8}$.

Substituting the equation to the inequality $y > 800$ yields $\frac{4x+5}{8} > 800$.

Multiplying 8 to both sides of the inequality yields $4x + 5 > 6{,}400$.

Subtracting 5 from both sides of the inequality yields $4x > 6{,}395$.

Dividing 4 from both sides of the equation yields $x > \frac{6{,}395}{4}$ or $x > 1{,}598.75$. Since x must be greater than 1,598.75, then the least integer value of x is 1,599.

72. **Level:** Hard | **Skill/Knowledge:** Linear inequalities in one or two variables | **Testing Point:** Solving linear inequalities

Key Explanation: Choice A is correct. Since her hair grows at least an inch every two months, then her hair grows at least half an inch every month. This means that $\textit{hair growth}_{\textit{after x months}} \geq 0.5x$. To create the inequality for the total length of her hair after x months, add the hair growth after x months to the length of the existing hair.

This yields $y \geq 0.5x + 10$.

Distractor Explanations: Choice B is incorrect and may result from not considering that her hair can grow more than 1 inch per two months. **Choice C** is incorrect and may result from an error in representing the statement "at least an inch every two months" into an expression. **Choice D** is incorrect and may result from calculating only the hair growth after x months and not considering the existing length.

73. **Level:** Hard | **Skill/Knowledge:** Linear inequalities in one or two variables | **Testing Point:** Combining linear inequalities in one variable to a linear inequality in two variables

Key Explanation: Choice C is correct. Rewriting the second inequality to have a ≥ symbol yields $30 \geq y$. Combining the two inequalities yields $x + 30 \geq 40 + y$. Subtracting y and 30 from both sides of the inequality yields $x - y \geq 40 - 30$ or $x - y \geq 10$.

Distractor Explanations: Choice A is incorrect and may result from a calculation error in combining the constants. **Choice B** is incorrect and may result from using the wrong inequality symbol. **Choice D** is incorrect and may result from using the wrong inequality symbol and sign for y.

74. **Level:** Hard | **Skill/Knowledge:** Linear inequalities in one or two variables | **Testing Point:** Interpreting compound inequalities

Key Explanation: 120 is the correct answer. The greatest possible value of the expression $x - y$ can be obtained using the maximum value of x and the minimum value of y. From the given inequalities, the maximum possible value of x is 90 and the minimum possible value of y is −30. Substituting the values to the expression $x - y$ yields $90 - (-30)$ or 120.

75. **Level:** Medium | **Skill/Knowledge:** Linear inequalities in one or two variables | **Testing Point:** Solve linear inequalities

Key Explanation: Choice D is the correct option. An absolute value inequality like $|A| \geq B$ is true when $A \geq B$ or $A \leq -B$.

Step 1: For the given inequality $|6x + 14| \geq 4x + 24$, we need to consider these two cases separately.

Case 1: $6x + 14 \geq 4x + 24$

Subtract $4x$ from both sides: $2x + 14 \geq 24$

Subtract 14 from both sides: $2x \geq 10$

Divide by 2: $x \geq 5$

Case 2: $6x + 14 \leq -(4x + 24)$

Simplify the right side: $6x + 14 \leq -4x - 24$

Add $4x$ to both sides: $10x + 14 \leq -24$

Subtract 14 from both sides: $10x \leq -38$

Divide by 10: $x \leq -3.8$

Step 2: Combine both conditions

We now have two conditions from the two cases:

$x \geq 5$ from the first case

$x \leq -3.8$ or $x \leq -\frac{19}{5}$ from the second case

These are the ranges of x that satisfy the inequality. Note that these are two separate intervals that do not overlap.

Distractor Explanations: Choices A and **B** are only part of the solution, and **Choice C** is the opposite of the solution.

76. **Level:** Medium | **Skill/Knowledge:** Linear inequalities in one or two variables | **Testing Point:** Solving linear inequalities

Key Explanation: 13 is the correct answer.

Given $\sqrt{2{,}016} < a\sqrt{14}$ dividing both sides of the inequality by $\sqrt{14}$ and putting the expression under one square root results in

$$\sqrt{\frac{2{,}016}{14}} < a.$$

This simplifies to: $\sqrt{144} < a$ or $12 < a$. So a has to be greater than 12.

Solving $\frac{a}{\sqrt{14}} < \sqrt{14}$ by multiplying both sides of the inequality by $\sqrt{14}$ results in $a < \sqrt{14}\sqrt{14}$.
This simplifies to $a < 14$.
So $12 < a < 14$.

Since a is an integer, $a = 13$.

77. **Level:** Medium | **Skill/Knowledge:** Linear inequalities in one or two variables | **Testing Point:** Solving linear inequalities

Key Explanation: Choice D is correct.
First Inequality:
$x + 2 \le -1$
To solve for x, subtract 2 from both sides:
$x \le -3$
Second Inequality:
$x > \frac{6}{3}x + 2$
This simplifies to:
$x > 2x + 2$
Subtract $2x$ from both sides to isolate x:
$-x > 2$
Multiply both sides by -1 (and reverse the inequality sign):
$x < -2$
Combining both inequalities:
$x \le -3$(from $x + 2 \le -1$)
$x < -2$(form $x > 2x + 2$)
Here, the range of x that satisfies both conditions is captured by the more restrictive condition:
$x \le -3$.

Distractor Explanations: Choice A is incorrect and is likely the result of not solving both systems of inequalities and combining the domains or from arithmetic errors. **Choice B** is incorrect and is likely the result of not solving both systems of inequalities and combining the domains or from arithmetic errors. **Choice C** is incorrect and is likely the result of not solving both systems of inequalities and combining the domains or from arithmetic errors.

78. **Level:** Easy | **Skill/Knowledge:** Linear inequalities in one or two variables | **Testing Point:** Solving linear inequalities

Key Explanation: Choice C is correct. In order to find the statement that is not true, choose a value for y which is greater than or equal to 4 and substitute in each option. Using 5 as the value of y and plugging it in **Choice C** yields $\frac{8}{5}(5) \ge 25$. Simplifying the inequality yields $8 \ge 25$ which is false. Therefore, **Choice C** is the correct answer.

Distractor Explanations: Choice A is true because it is exactly the same with $y \ge 4$. **Choice B** is also true because $y \ge 4$ makes $4y \ge 16$, therefore $4y + 7 \ge 16 + 7 > 22$. **Choice D** is true because $y \ge 4$ makes $\frac{1}{2}y \ge 2$, therefore $\frac{1}{2}y - \frac{1}{2} \ge 0 \ge -\frac{1}{2}$.

79. **Level:** Easy | **Skill/Knowledge:** Linear inequalities in one or two variables | **Testing Point:** Solving linear inequalities

Key Explanation: Choice B is correct. The inequality must be less than or equal to 425 feet of yarn, as that is the maximum value Jane purchased. The values 15 and 12 must be the coefficients of the variables (s) and (h), respectively.

Since the total yarn to be used is represented by the expression $15s + 12h$, then the inequality is $425 \geq 15s + 12h$.

Distractor Explanations: Choice A is incorrect and has interchanged the amount of yarn required for scarf and hat. **Choice C** is incorrect and has interchanged the amount of yarn required for scarf. **Choice D** is incorrect as it shows that total yarn used is greater than the yarn purchased.

80. **Level:** Easy | **Skill/Knowledge:** Linear inequalities in one or two variables | **Testing Point:** Solving linear inequalities

Key Explanation: Choice B is correct. If the frame can hold up to 12 pounds, the variable (s) must be greater than or equal to the sum of the other variables. Since the total weight of the painting and the frame is $x + y$, then the inequality is $s \geq x + y$ or $s \geq y + x$. Therefore, **Choice B** is correct.

Distractor Explanations: Choice A is incorrect and is the result of errors in creating inequalities. **Choice C** is incorrect and is the result of errors in creating inequalities. **Choice D** is incorrect and is the result of errors in creating inequalities.

81. **Level:** Medium | **Skill/Knowledge:** Linear inequalities in one or two variables | **Testing Point:** Solving linear inequalities with absolute value

Key Explanation: Choice C is correct. To find the possible values of t, make two inequalities using the content of the absolute value and the positive and negative values of the constant.

This yields $2t - 2 < 1$ and $2t - 2 > -1$. Solving the values of t yields $t < \frac{3}{2}$ and $t > \frac{1}{2}$. Combining the two inequalities yields $\frac{1}{2} < t < \frac{3}{2}$ or $0.5 < t < 1.5$. Since t is an integer, then t can only be equal to 1.

Distractor Explanations: Choice A is incorrect and may result from errors in evaluating absolute value inequalities. **Choice B** is incorrect and may result from errors in evaluating absolute value inequalities. **Choice D** is incorrect and may result from errors in evaluating absolute value inequalities.

82. **Level:** Easy | **Skill/Knowledge:** Linear inequalities in one or two variables | **Testing Point:** Solving linear inequalities

Key Explanation: Choice B is correct. The greatest possible integer value of p can be obtained using the least value of q. Since q is a positive multiple of 6, the least value of q is 6. Substituting 6 for q in the given inequality yields $p(6) < 100$. Dividing both sides of the inequality by 6 yields $p < 16.67$. Therefore, the greatest possible integer value of p is 16.

Distractor Explanations: Choice A is incorrect and may result from finding the least value of q. **Choice C** is incorrect and may result from a conceptual or calculation error. **Choice D** is incorrect and may result from finding the value of pq when $p = 16$ and $q = 6$.

83. **Level:** Hard | **Skill/Knowledge:** Linear inequalities in one or two variables | **Testing Point:** Solving compound linear inequalities

Key Explanation: Choice A is correct. To find the range of the possible values of the expression

$r - s$, subtract the opposite boundary limits of the two variables.

To get the minimum boundary of $r - s$, subtract the maximum boundary of s from the minimum boundary of r which yields $20 - 60 = -40$.

To get the maximum boundary of $r - s$, subtract the minimum boundary of s from the maximum boundary of r which yields $30 - 40 = -10$.

Therefore, the correct answer is $-40 < r - s < -10$.

Distractor Explanations: Choice B is incorrect and reflects error in interpreting inequalities. **Choice C** is incorrect and reflects error in interpreting inequalities. **Choice D** is incorrect and reflects error in interpreting inequalities.

84. **Level:** Medium | **Skill/Knowledge:** Linear inequalities in one or two variables | **Testing Point:** Solving for systems of linear inequalities

Key Explanation: Choice A is correct. To determine if a point is a solution, plug it into both inequalities and verify if the statements are true. Plugging in (2, 4) from **Choice A** to the 1st inequality yields $2(2) + 5 \leq 3(2) + 7$. Simplifying it yields $9 \leq 13$, which is true.

Plugging in (2, 4) to the 2nd inequality yields $2(4) > -(2) + 4$. Simplifying it yields $8 > 2$, which is also true. Since both statements are true, then (2, 4) is a solution to the system of inequalities.

Distractor Explanations: Choice B is incorrect. Substituting (−4, 3) to the 2nd inequality yields $2(3) > -(-4) + 4$ or $6 > 8$, which is false. **Choice C** is incorrect. Substituting (2, 1) to the 2nd inequality yields $2(1) > -2 + 4$ or $2 > 2$, which is false. **Choice D** is incorrect. Substituting (−3, 3) to the 2nd inequality yields $2(3) > -(-3) + 4$ or $6 > 7$, which is false.

85. **Level:** Easy | **Skill/Knowledge:** Linear inequalities in one or two variables | **Testing Point:** Solving linear inequality equations

Key Explanation: Choice C is correct. The baker can bake a maximum of 4 loaves of bread a day. The least number of loaves of bread that the baker can make is 1. Hence, the maximum number of cakes the baker can make is $27 - 1 = 26$ cakes.

Distractor Explanations: Choice A is incorrect. This is the value of the least number of cakes they can bake in a day. **Choice B** is incorrect. This option can be a result of a conceptual error. **Choice D** is incorrect. This option can be a result of a conceptual error.

86. **Level:** Easy | **Skill/Knowledge:** Linear inequalities in one or two variables | **Testing Point:** Solving linear inequalities

Key Explanation: The correct answer is 4. Since $12n < 49$, $n < \frac{49}{12} = 4.1$. Since $7n > 22$, then $n > \frac{22}{7} = 3.1$. Combining the inequalities yields $3.1 < n < 4.1$.

Therefore, n can only be 4.

87. **Level:** Easy | **Skill/Knowledge:** Linear inequalities in one or two variables | **Testing Point:** Creating and solving equations for linear inequalities

Key Explanation: Choice B is correct. Since Carmen buys only one box of pens, the amount left that she can only spend to buy books is \$43 − \$5 = \$38. Let b be the number of books that she will buy. Hence, $3b$ is the total amount of money to be spent on books. Since she can only spend up to \$38, then $3b \leq 38$. Dividing both sides of the inequality by 3 yields $b \leq 12.67$. The next lower integer is 12. Therefore, Carmen can buy a maximum of 12 books.

Distractor Explanations: Choice A is incorrect. This is the cost per book. **Choice C** is incorrect. A book costs $3, however, after buying 12 books and 1 box of pens, she is left with $2, which is not enough for the 13th book. **Choice D** is incorrect. This is the amount she can spend after buying a box of pens.

Chapter 4

Advanced Math

This chapter includes questions on the following topics:

- Equivalent expressions
- Nonlinear equations in one variable and systems of equations in two variables
- Nonlinear functions

88

Which answer correctly simplifies the expression $50x^2 + 40x - 150 - 20(x^2 + x - 1)$?

A) $30x^2 + 60x - 170$

B) $70x^2 - 60x - 170$

C) $30x^2 + 50x - 180$

D) $30x^2 + 20x - 130$

89

Simplify the following expression:
$x(x - 5) + 4(x - 3) - 8$.

A) $(x - 10)(x + 2)$

B) $(x - 4)(x + 3)$

C) $(x - 5)(x + 4)$

D) $x^2 - x - 8$

90

Which of the following expressions is equal to $\frac{x^{-1}y^2}{y^3x^{-2}}$?

A) $\frac{y}{x}$

B) x

C) y

D) $\frac{x}{y}$

91

Simplify the following expression:

$\frac{2x-1}{x-3} - \frac{x-1}{x+2}$.

A) $\frac{x^2+7x-5}{x^2-x-6}$

B) $\frac{x^2+7x-5}{x^2+x-6}$

C) $\frac{x^2-x-1}{2x-1}$

D) $\frac{x^2+x-1}{(x+2)(x-3)}$

92

The difference of $-7x^2 + 4x - 12$ and $3x^2 + 16x + 4$ can be written in the form $ax^2 + bx + c$ where a, b, and c are constants. What is the value of $a \times c$?

93

What is the value of m in the equation $\left(2^{m+n}\right)^{m-n} = 128$ if $n^2 = 9$ and if $m > 0$?

94

The equation of the acceleration of an object sinking in water can be represented by

$$a = g\left(1 - \left(\frac{m_w}{m_0}\right)\right)$$

where a is the acceleration in meters per second, g is the gravitational constant of 9.8 meters per second squared, m_w is the mass of the displaced water in grams, and m_0 is the mass of the object itself in grams.

Which of the following equations expresses the mass of the displaced water (m_w) in terms of the other variables?

A) $m_w = -m_0\left(\left(\frac{a}{g}\right) - 1\right)$

B) $m_w = -\frac{m_0}{\left(\frac{a}{g}\right) - 1}$

C) $m_w = -m_0\left(\left(\frac{a}{g}\right) + 1\right)$

D) $m_w = \left(\frac{am_0}{g} + 1\right)$

95

What is the difference of $(7 - 2x) - (-8x + 10x^2)$?

A) $10x^2 - 2x + 15$

B) $-10x^2 + 6x + 7$

C) $10x^2 - 2x - 1$

D) $-10x^2 - 2x - 1$

96

$$(a^2b - 3a^2 + 10ab^2) - (-2a^2b - 2ab^2 - 3b^2)$$

Which of the following is equivalent to the preceding expression?

A) $3a^2b - 3a^2 + 3b^2 - 12ab^2$

B) $3a^2b - 12ab^2$

C) $3a^2b - 3a^2 + 3b^2 + 12ab^2$

D) $15a^2b^2 + 3a^2 + 3b^2$

97

If $m > 4$, which of the following is equivalent to

$$\frac{1}{\left(\frac{1}{m+2}\right) + \left(\frac{1}{m+4}\right)}?$$

A) $\frac{1}{2m+6}$

B) $2m + 6$

C) $m^2 + 6m + 8$

D) $\frac{m^2 + 6m + 8}{2m + 6}$

98

If a hours and 30 minutes is equal to $4\frac{1}{2}$ hours, what is the value of a in quarter hours?

99

Simplify the following expression:
$y(y-5)+4(y-4)-4$.

A) $(y+11)^2$

B) $11y-11$

C) $(y-5)(y+4)$

D) y^2-y-8

100

Which of the following expressions is equal to
$\frac{a^{-2}b^4}{b^6a^{-4}}$?

A) $\frac{b^2}{a^2}$

B) a^2

C) b^2

D) $\frac{a^2}{b^2}$

101

Which of the following is equivalent to $6x^2(xyz)$?

A) $6x^3 6xy\ 6xz$

B) $216xyz^2$

C) $6x^3yz$

D) $36x^3yz$

102

For all values of $z>0$, which of the following expressions is equivalent to

$$\frac{\frac{1}{x}+\frac{1}{y}}{z}?$$

A) $\frac{y+x}{z}$

B) $\frac{zy+zx}{xyz}$

C) $\frac{xy}{z(x+y)}$

D) $\frac{y+x}{zxy}$

103

If $\frac{2x^2+6x-11}{x-3}=2x+12+\frac{B}{x-3}$, what is the value of B?

A) -11

B) 14

C) 25

D) 36

104

If $5(xy+yz)-2(xz-xy)=axy+byz+cxz$, what is the value of a?

A) -2

B) 5

C) 7

D) 10

Equivalent expressions

105

If the expression $a^{\frac{3}{2}} \cdot b^{\frac{1}{3}}$ is equivalent to $\sqrt[p]{(a)^{11}}$, what is the value of p if $b = a$?

106

If $ax^2 + bx + c = (2x + 1)(x - 5) - 13$, then what is the value of $a + b + c$?

107

Which of the following expresses the parabola $y = (x + 3)(x - 1)$ in vertex form?

A) $y = x^2 + 2x - 3$

B) $y = (x - 1)^2 + 4$

C) $y = (x + 1)^2 - 4$

D) $y = x^2 - 2x + 3$

108

A parabolic function can be written in the terms of $f(x) = ax^2 + bx + c$. What is the value of $-c$ after simplifying the function $f(x) = (7x^2 + 3x - 18) - (-3x^2 - 17x + 19)$?

A) -41

B) -37

C) 37

D) 41

88. **Level:** Medium | **Skill/Knowledge:** Equivalent expressions | **Testing Point:** Simplifying quadratic expressions by using the distributive property and combining like terms

Key Explanation: Choice D is correct. Using the distributive property yields $50x^2 + 40x - 150 - 20x^2 - 20x + 20$. Combining like terms yields $30x^2 + 20x - 130$.

Distractor Explanations: Choice A is incorrect and may result from an error in distributing the negative sign inside the parentheses. **Choice B** is incorrect and may result from a conceptual or calculation error. **Choice C** is incorrect and may result from a conceptual or calculation error.

89. **Level:** Medium | **Skill/Knowledge:** Equivalent expressions | **Testing Point:** Using the distributive property to simplify a linear expression

Key Explanation: Choice C is correct. To simplify the expression, use distributive property which yields $x^2 - 5x + 4x - 12 - 8$. Combining like terms yields $x^2 - x - 20$.

The equation can be simplified once more by factoring which yields $(x - 5)(x + 4)$.

Distractor Explanations: Choice A is incorrect and may result from an error in factoring $x^2 - x - 20$. **Choice B** is incorrect and may result from a calculation error which leads to the wrong expression $x^2 - x - 12$. **Choice D** is incorrect and may result from disregarding -12 in the expression.

90. **Level:** Medium | **Skill/Knowledge:** Equivalent expressions | **Testing Point:** Using exponent rules to rewrite a rational expression

Key Explanation: Choice D is correct. Start by simplifying the exponents of each variable. Subtracting the exponents of the matching bases being divided yields $x^{-1-(-2)}y^{2-3}$. Simplifying the expression yields xy^{-1} and according to the inverse exponent rules, the expression can be rewritten as $\frac{x}{y}$.

Distractor Explanations: Choice A is incorrect and may result from a conceptual or calculation error. **Choice B** is incorrect and may result from a conceptual or calculation error. **Choice C** is incorrect and may result from a conceptual or calculation error.

91. **Level:** Medium | **Skill/Knowledge:** Equivalent expressions | **Testing Point:** Combining rational expressions through the least common denominator

Key Explanation: Choice A is correct. To simplify the expression, rewrite the fractions using the least common denominator which is $(x - 3)(x + 2)$. This yields $\frac{(2x-1)(x+2)}{(x-3)(x+2)} - \frac{(x-1)(x-3)}{(x+2)(x-3)}$ or $\frac{(2x-1)(x+2)-(x-3)(x-1)}{(x-3)(x+2)}$. Using distributive property yields $\frac{2x^2-x+4x-2-x^2+4x-3}{x^2-x-6}$.

Combining like terms yields $\frac{x^2+7x-5}{x^2-x-6}$.

Distractor Explanations: Choice B is incorrect and may result from calculation error in simplifying polynomials and fractions. **Choice C** is incorrect and may result from calculation error in simplifying polynomials and fractions. **Choice D** is incorrect and may result from calculation error in simplifying polynomials and fractions.

92. **Level:** Medium | **Skill/Knowledge:** Equivalent expressions | **Testing Point:** Operations with quadratic equations

Key Explanation: The correct answer is 160. Subtracting the second polynomial from the first polynomial and combining like terms results in
$-7x^2 + 4x - 12 - (3x^2 + 16x + 4)$
$= -7x^2 + 4x - 12 + (-3x^2) + (-16x) + (-4)$
$= -10x^2 - 12x - 16$.

The standard form of a quadratic equation is $y = ax^2 + bx + c$. Therefore, $a = -10$, $b = -12$, $c = -16$. The product of a and c is $(-10) \times (-16) = 160$.

93. **Level:** Hard | **Skill/Knowledge:** Equivalent expressions | **Testing Point:** Using exponent rules and the difference of two squares

Key Explanation: Using exponent rules to raise a power to a power results in the equation being rewritten as: $(2)^{(m+n)(m-n)} = 128$. $(m + n)(m - n)$ represents the factored form of the difference of two squares and can be written as $m^2 - n^2$. Thus, $2^{m\wedge 2 - n\wedge 2} = 128$. $2^7 = 128$, so $2^7 = 2^{m\wedge 2 - n\wedge 2}$. Since $n^2 = 9$, substituting 9 for n^2 in the equation and solving for m results in
$2^7 = 2^{m\wedge 2 - n\wedge 2}$
$2^7 = 2^{m\wedge 2 - 9}$.

Since the bases are the same, the exponents must also be equal. So:
$m^2 - 9 = 7$. Adding 9 to both sides yields $m^2 = 16$. Taking the square root of both sides results in $m = \pm 4$.

Since $m > 0$, the answer is 4.

94. **Level:** Easy | **Skill/Knowledge:** Equivalent expressions | **Testing Point:** Rewriting a rational expression to solve for a particular variable

Key Explanation: Given the equation

$$a = g\left(1 - \left(\frac{m_w}{m_0}\right)\right).$$

To solve for m_w, first divide both sides of the equation by g to get:

$1 - \left(\frac{m_w}{m_0}\right) = \frac{a}{g}$. Next, subtract 1 from both sides of the equation to get:

$-\frac{m_w}{m_0} = \left(\frac{a}{g}\right) - 1$. Then multiply both sides of the equation by $-m_0$ to get:

$m_w = -m_0\left(\left(\frac{a}{g}\right) - 1\right)$, which is **Choice A.**

Distractor Explanations: Choice B would be incorrectly picked if a mistake was made in isolating the correct variable. **Choice C** would be selected if a math calculation mistake was made. **Choice D** would be incorrectly picked if a mistake was made in isolating the correct variable.

95. **Level:** Easy | **Skill/ Knowledge:** Equivalent expressions | **Testing Point:** Using the distributive property and combining like terms

Key Explanation: Choice B is correct.

Using distributive property to simplify the expression $(7 - 2x) - (-8x + 10x^2)$ yields $7 - 2x + 8x - 10x^2$.

Combining like terms yields $-10x^2 + 6x + 7$.

Distractor Explanations: Choice A is incorrect and may result from a conceptual or calculation error. **Choice C** is incorrect and may result from a conceptual or calculation error. **Choice D** is incorrect and may result from a conceptual or calculation error.

96. **Level:** Easy | **Skill/Knowledge:** Equivalent expressions | **Testing Point:** Using the distributive property and combining like terms

Key Explanation: Choice C is correct. Start by distributing the negative sign to the parentheses of the second term in the expression $(a^2b - 3a^2 + 10ab^2) - (-2a^2b - 2ab^2 - 3b^2)$.

This yields $a^2b - 3a^2 + 10ab^2 + 2a^2b + 2ab^2 + 3b^2$.

Combining like terms yields $3a^2b - 3a^2 + 3b^2 + 12ab^2$.

Distractor Explanations: Choice A is incorrect and reflects errors in simplifying complex polynomials. **Choice B** is incorrect and reflects errors in simplifying complex polynomials. **Choice D** is incorrect and reflects errors in simplifying complex polynomials.

97. **Level:** Hard | **Skill/Knowledge:** Equivalent expressions | **Testing Point:** Rewriting a rational expression by finding the least common denominator

Key Explanation: Choice D is correct.

Firstly, find the *LCM* for the denominators of $m + 2$ and $m + 4$. The *LCM* is $(m + 2)(m + 4)$.

Simplifying the denominator yields

$$\frac{1}{\left(\frac{m+4+m+2}{(m+2)(m+4)}\right)} \text{ or } \frac{1}{\frac{2m+6}{m^2+6m+8}}.$$

The denominator in the denominator will become the numerator which yields

$$\frac{m^2+6m+8}{2m+6}.$$

Distractor Explanations: Choice A is incorrect and reflects error in simplifying complex fractions with different denominators. **Choice B** is incorrect and reflects error in simplifying complex fractions with different denominators. **Choice C** is incorrect and reflects error in simplifying complex fractions with different denominators.

98. **Level:** Easy | **Skill/Knowledge:** Equivalent expressions | **Testing Point:** Converting between units

Key Explanation: The correct answer is 16. Converting 30 minutes to hours yields $\frac{30}{60}$ or $\frac{1}{2}$ hour. Since $a + \frac{1}{2}$ hour $= 4\frac{1}{2}$ hours, then $a = 4$ hours. Converting 4 hours to quarter hours yields $4(4) = 16$ quarter hours.

99. **Level:** Easy | **Skill/Knowledge:** Equivalent expressions | **Testing Point:** Using the distributive property to simplify a quadratic expression

Key Explanation: Choice C is correct. In order to simplify the expression, distribute and combine like terms which yields $y^2 - 5y + 4y - 16 - 4$ or $y^2 - y - 20$.

The equation can be simplified once more by factoring which yields $(y - 5)(y + 4)$.

Therefore, the correct answer is **Choice C.**

Distractor Explanations: Choice A is incorrect and is the result of incorrectly distributing values, combining like terms, or factoring polynomials. **Choice B** is incorrect and is the result of incorrectly distributing values, combining like terms, or factoring polynomials. **Choice D** is incorrect and is the result of incorrectly distributing values, combining like terms, or factoring polynomials.

Equivalent expressions (Answers)

100. **Level:** Easy | **Skill/Knowledge:** Equivalent expressions | **Testing Point:** Using exponent rules to rewrite a rational expression

Key Explanation: Choice D is correct. Start by simplifying the exponents of each variable. By subtracting the exponents of the matching bases being divided yields $a^{-2-(-4)}\ b^{4-6}$. This can be simplified to a^2b^{-2}. Since b has a negative exponent, it will be moved to the denominator to make the exponent positive. This yields $\frac{a^2}{b^2}$.

Distractor Explanations: Choice A is incorrect and may result from a conceptual or calculation error. **Choice B** is incorrect and may result from a conceptual or calculation error. **Choice C** is incorrect and may result from a conceptual or calculation error.

101. **Level:** Easy | **Skill/Knowledge:** Equivalent expressions | **Testing Point:** Using the distributive property

Key Explanation: Choice C is correct. Using distributive property to simplify $6x^2(xyz)$ yields $6x^3yz$.

Distractor Explanations: Choice A is incorrect and may result from conceptual error. **Choice B** is incorrect and may result from conceptual error. **Choice D** is incorrect and may result from conceptual error.

102. **Level:** Hard | **Skill/Knowledge:** Equivalent expressions | **Testing Point:** Adding rational expressions

Key Explanation: Choice D is correct. Simplifying the numerator first yields $\frac{1}{x}+\frac{1}{y}=\frac{y+x}{xy}$.

This would result to $\frac{\frac{y+x}{xy}}{z}$. By multiplying both the numerator and denominator by $\frac{1}{z}$, we get $\frac{y+x}{xyz}$, which is **Choice D**.

Distractor Explanations: Choice A is incorrect and may be a result of conceptual or calculation error. **Choice B** is incorrect and may be a result of conceptual or calculation error. **Choice C** is incorrect and may be a result of conceptual or calculation error.

103. **Level:** Easy | **Skill/Knowledge:** Equivalent expressions | **Testing Point:** Using the remainder theorem

Key Explanation: Choice C is correct. B is the value of the remainder when $2x^2 + 6x - 11$ is divided by $x - 3$. The remainder would be the value of $f(3)$.

Substituting 3 to x in $2x^2 + 6x - 11$ yields $2(3)^2 + 6(3) - 11$ or 25.

Distractor Explanations: Choice A is incorrect. This is the value of the y-intercept of the equation $y = 2x^2 + 6x - 11$. **Choice B** is incorrect. This may be a result of a miscalculation or conceptual error. **Choice D** is incorrect. This may be a result of a miscalculation or conceptual error.

104. **Level:** Easy | **Skill/Knowledge:** Equivalent expressions | **Testing Point:** Matching coefficients

Key Explanation: Choice C is correct. Using the distributive property to simplify $5(xy + yz) - 2(xz - xy)$ yields $5xy + 5yz - 2xz + 2xy$. Combining like terms yields $7xy + 5yz - 2xz$. Matching the coefficients to $axy + byz + cxz$ yields $a = 7$, $b = 5$, and $c = -2$.

Distractor Explanations: Choice A is incorrect. This is the value of c which is coefficient of xz. **Choice B** is incorrect. This is the value of b which is coefficient of yz. **Choice D** is incorrect. This is the value of $a + b + c$.

105. **Level:** Hard | **Skill/Knowledge:** Equivalent expressions | **Testing Point:** Converting from exponents to radicals

Key Explanation: 6 is the correct answer. If $b = a$, then $a^{\frac{3}{2}} \cdot b^{\frac{1}{3}} = a^{\frac{3}{2}} \cdot b^{\frac{1}{3}}$. Since the bases are the same, the exponents should therefore be added.

Adding the exponents yields $\frac{3}{2} + \frac{1}{3} = \frac{9+2}{6} = \frac{11}{6}$.

Equating the two expressions yields $a^{\frac{11}{6}} = \sqrt[p]{a^{11}}$. Rewriting the right side of the equation in exponential form yields $a^{\frac{11}{6}} = a^{\frac{11}{p}}$. Matching the exponents yields $p = 6$.

106. **Level:** Medium | **Skill/Knowledge:** Equivalent expressions | **Testing Point:** Foiling out binomials and matching coefficients

Key Explanation: -25 is the correct answer. To determine the values of a, b, c, simplify the right side of the equation by using distributive property which yields $2x^2 - 10x + x - 5 - 13$. Combining like terms yields $2x^2 - 9x - 18$. Comparing the coefficients and constant on the equation $ax^2 + bx + c = 2x^2 - 9x - 18$ yields $a = 2$, $b = -9$ and $c = -18$. Substituting the value to the expression $a + b + c$ yields $2 - 9 - 18$. Therefore, the value of the expression is -25.

107. **Level:** Medium | **Skill/Knowledge:** Equivalent expressions | **Testing Point:** Converting from one form of a parabola to another

Key Explanation: Choice C is correct. The equation of a parabola in vertex form is $y = (x - h)^2 + k$. To rewrite the given equation into vertex form, expand the right side of the equation which yields $y = x^2 + 2x - 3$. Completing the square and grouping the perfect square trinomial yields $y = (x^2 + 2x + 1) - 4$. Simplifying the equation yields $y = (x + 1)^2 - 4$.

Distractor Explanations: Choice A is incorrect and is the equation in the expanded form. **Choice B** is incorrect and may be due to error in signs while expanding or converting. **Choice D** is incorrect and may be due to error in signs while expanding the equation.

108. **Level:** Easy | **Skill/Knowledge:** Equivalent expressions | **Testing Point:** Using the distributive property and combining like terms

Key Explanation: The correct answer is **Choice C**. Using distributive property to simplify the expression $(7x^2 + 3x - 18) - (-3x^2 - 17x + 19)$ yields $7x^2 + 3x - 18 + 3x^2 + 17x - 19$. Combining like terms yields $10x^2 + 20x - 37$ where $c = -37$. Therefore, $-c = 37$.

Distractor Explanations: Choice A is incorrect and may result from a calculation error. **Choice B** is incorrect and may result from solving only the value of c. **Choice D** is incorrect and may result from a calculation error.

109

If $\frac{x+4}{3}=2$, then what is the value of $\frac{x+2}{(x-1)^2}$?

A) 0

B) 2

C) 3

D) 4

110

If $m+4=\sqrt{m+10}$, what is the solution set of the equation?

A) {−6, 1}

B) {−6}

C) {−1}

D) {6}

111

$$y = x^3 + 4$$
$$y(x - 4) = x^2$$

How many solutions are there at most to the system of equations above?

A) 1 solution

B) 2 solutions

C) 3 solutions

D) 4 solutions

112

$$x^2 + x - 90 = 0$$

If c is a solution to the equation above, and $c > 0$, what is the value of c?

113

$$2^x \times 2^y = 1{,}024$$

In the equation above, what is the value of $x + y$?

114

If $(px + 3)(qx + 8) = 6x^2 + rx + 24$ for all values of x and $p + q = 5$, what are the two possible values of p?

A) 2 and 3

B) 6 and 1

C) 8 and 2

D) −1 and −6

115

If $d > 0$ and $d^2 - 16 = 0$, what is the value of d?

116

If $x = 7\sqrt{2}$ and $2x = \sqrt{2y}$, what is the value of y?

117

If $a + b = p + rq$ and $q \neq 0$, then r is represented by which expression?

A) $\frac{a+b+p}{q}$

B) $a-b+\frac{p}{q}$

C) $a+b+\frac{p}{q}$

D) $\frac{a+b-p}{q}$

118

If k is a positive integer and $2k^3 - 2k = 0$, what is the value of k?

119

$$|-2x + 6| = -4$$

How many solutions does the equation above have?

A) None

B) One

C) Two

D) Infinite

120

$$4x^2 - 17x + 4 = 0$$

If m and n are solutions to the quadratic equation above, what is the value of $m + n$?

A) $\frac{1}{4}$

B) 1

C) 4

D) $\frac{17}{4}$

121

$$3x^2 - bx - 6 = 0$$

If one of the solutions for the equation above is $-\frac{1}{3}$, what is the value of the other roots?

122

If $b^2 - \frac{1}{4} = 0$ and the values of b are real numbers, what is the sum of all possible values of b?

123

$$3(x - 2)^2 + 5 = 3x^2 + 12x - 17$$

How many solutions does the equation above have?

A) None

B) One

C) Two

D) Infinite

109. **Level:** Easy | **Skill/Knowledge:** Nonlinear equations in one variable and systems of equations in two variables | **Testing Point:** Solving for a variable in an equation and finding the correct value by substitution

Key Explanation: Choice D is correct. Use the equation $\frac{x+4}{3}=2$ to get the value of x. Multiplying both sides of the equation by 3 yields $x + 4 = 6$. Subtracting 4 from both sides of the equation yields $x = 2$. Substituting $x = 2$ in the expression $\frac{x+2}{(x-1)^2}$ yields $\frac{2+2}{(2-1)^2}=\frac{4}{1}=4$.

Distractor Explanations: Choice A is incorrect and may result due to finding $x - 2$. **Choice B** is incorrect and is the value of x. **Choice C** is incorrect and may result from a conceptual or calculation error.

110. **Level:** Medium | **Skill/Knowledge:** Nonlinear equations in one variable and systems of equations in two variables | **Testing Point:** Solving equations with square roots in one variable

Key Explanation: Choice C is correct. Start by squaring both sides of the equation to remove the square root.
The left side of the equation becomes $(m + 4)^2 = m^2 + 8m + 16$.
The right side of the equation becomes $m + 10$.
Hence, the equation becomes
$m^2 + 8m + 16 = m + 10$.
Subtracting m and 10 from both sides of the equation yields
$m^2 + 7m + 6 = 0$. Factoring the quadratic equation results in
$(m + 6)(m + 1) = 0$.

Setting the factors equal to zero results in the solutions $m = -1$ or $m = -6$.

When sides of an equation are squared, there can be extraneous solutions created. Thus, both -6 and -1 need to be checked into the original equation to see if there are any extraneous solutions. When -6 is substituted for m in the original equation, the result is $-6 + 4 = \sqrt{-6+10}$. Simplifying it yields $-2 = 2$. Since the statement is false, the -6 is not valid.

When -1 is substituted for m in the original equation, the result is $-1+ 4 = \sqrt{-1+10}$. Simplifying the equation yields $3 = 3$. Since the statement is true, then -1 is a valid solution. Therefore, $m = -1$.

Distractor Explanations: Choice A is incorrect. This is the result of arithmetic error and not checking the answers into the original equation to eliminate the extraneous solution. **Choice B** is incorrect. This is the result of arithmetic error and not checking the answers into the original equation to eliminate the extraneous solution. **Choice D** is incorrect. This is the result of arithmetic error and not checking the answers into the original equation to eliminate the extraneous solution.

111. **Level:** Easy | **Skill/Knowledge:** Nonlinear equations in one variable and systems of equations in two variables | **Testing Point:** Solving systems of nonlinear equations in two variables

Key Explanation: Choice D is correct. Substituting the value of y from the first equation into the second equation results in $(x^3 + 4)(x - 4) = x^2$.

Multiplying out the binomials on the left side of the equation yields

$$x^4 - 4x^3 + 4x - 16 = x^2.$$

Subtracting x^2 from both sides of the equation results in

$$x^4 - 4x^3 - x^2 + 4x - 16 = 0.$$

The n-Roots Theorem states $f(x)$ is a polynomial of degree n, where $n \neq 0$, then $f(x)$ has at most n zeros. Therefore, according to this theorem, since the above polynomial is of degree 4, there are at most 4 solutions to the above equation.

Distractor Explanations: Choice A is incorrect and is either the result of lack of knowledge of the n-Roots Theorem or result of arithmetic errors. **Choice B** is incorrect and is either the result of lack of knowledge of the n-Roots Theorem or result of arithmetic errors. **Choice C** is incorrect and is either the result of lack of knowledge of the n-Roots Theorem or result of arithmetic errors.

112. **Level:** Easy | **Skill/Knowledge:** Nonlinear equations in one variable and systems of equations in two variables | **Testing Point:** Solving a quadratic equation

Key Explanation: The correct answer is 9. The equation $x^2 + x - 90 = 0$ can be factored as follows:
$(x + 10)(x - 9) = 0$. Setting both terms equal to zero yields

$$x = -10 \text{ or } 9.$$

Since $c > 0$, the value of c is 9 as it is the only solution greater than zero.

113. **Level:** Easy | **Skill/Knowledge:** Nonlinear equations in one variable and systems of equations in two variables | **Testing Point:** Using rules of exponents to solve equation

Key Explanation: The value of $x + y$ is 10. Using exponent rules for multiplication with the same base, find that $2^x \times 2^y = 2^{x+y} = 1{,}024$. Since, $2^{10} = 1{,}024$, $2^{x+y} = 2^{10}$ and thus $x + y = 10$.

114. **Level:** Easy | **Skill/Knowledge:** Nonlinear equations in one variable and systems of equations in two variables | **Testing Point:** Solving systems of linear and nonlinear equations in two variables

Key Explanation: Choice A is correct. Using distributive property to simplify the expression, $(px + 3)(qx+ 8)$ yields $(pq)\, x^2 + 8px + 3qx + 24$. Combining like terms makes the equation $(pq)\, x^2 + (8p + 3q)x + 24 = 6x^2 + rx + 24$. Matching the coefficient of x^2 on both sides of the equation yields $pq = 6$. Since $p + q = 5$, then the factors of 6 that satisfy the equation must be 2 and 3. Therefore, the possible values of p are 2 and 3.

Distractor Explanations: Choice B is incorrect and reflects error in factoring polynomials. **Choice C** is incorrect and reflects error in factoring polynomials. **Choice D** is incorrect and reflects error in factoring polynomials.

115. **Level:** Easy | **Skill/Knowledge:** Nonlinear equations in one variable and systems of equations in two variables | **Testing Point:** Solving a quadratic equation by the square root method

Key Explanation: The correct answer is 4. Since $d^2 - 16 = 0$, then $d^2 = 16$. Solving for d by getting the square root of both sides of the equation yields $d = \pm 4$.

The question states $d > 0$, therefore, the answer is 4.

116. **Level:** Easy | **Skill/Knowledge:** Nonlinear equations in one variable and systems of

equations in two variables | **Testing Point:** Solving systems of nonlinear equations in two variables with square roots

Key Explanation: The correct answer is 196. To find the value of y, substitute the value of x from the 1st equation to the 2nd equation.

This yields $2\times7\sqrt{2}=\sqrt{2y}$ or $14\sqrt{2}=\sqrt{2y}$.

Grouping the radical signs yields $14\times\sqrt{2}=\sqrt{2}\times\sqrt{y}$.

Dividing both sides of the equation by $\sqrt{2}$ yields $14=\sqrt{y}$.

Squaring both sides of the equation yields $y = 196$.

117. **Level:** Easy | **Skill/Knowledge:** Nonlinear equations in one variable and systems of equations in two variables | **Testing Point:** Solving for one variable in terms of another

Key Explanation: Choice D is correct. Since $a + b = p + rq$, solve for r by isolating it.

Subtracting p from both sides of the equation yields $a + b - p = rq$.

Dividing both sides of the equation by q yields $\frac{a+b-p}{q}=r$.

Distractor Explanations: Choice A is incorrect and reflects error in analyzing multiple equations with the same variables. **Choice B** is incorrect and reflects error in analyzing multiple equations with the same variables. **Choice C** is incorrect and reflects error in analyzing multiple equations with the same variables.

118. **Level:** Medium | **Skill/Knowledge:** Nonlinear equations in one variable and systems of equations in two variables | **Testing Point:** Solving nonlinear equation in one variable

Key Explanation: The correct answer is 1.

To find k, factor the given equation.

Factoring out $2k$ yields $2k(k^2 - 1) = 0$.

Factoring the difference of two squares binomial yields $2k(k + 1)(k - 1)=0$.

Equating each factor to zero and finding the value of k yields $k = 0$, $k = -1$ and $k = 1$. *i.e.* $k = \{-1, 0, 1\}$.

Since k is a positive integer, then $k = 1$.

119. **Level:** Medium | **Skill/Knowledge:** Nonlinear equations in one variable and systems of equations in two variables | **Testing Point:** Finding solutions for absolute value equations

Key Explanation: Choice A is correct. Absolute values are always positive. An absolute value equal to a negative value will have no solutions.

Distractor Explanations: Choice B is incorrect. For an absolute value to have 1 solution, it should be equal to 0. **Choice C** is incorrect. For an absolute value to have two solutions, it should be equal to a positive number. **Choice D** is incorrect. Equations involving absolute values cannot have infinite solutions.

120. **Level:** Easy | **Skill/Knowledge:** Nonlinear equations in one variable and systems of equations in two variables | **Testing Point:** Finding the sum of the solutions in a quadratic equation

Key Explanation: Choice D is correct. The sum of solutions in a quadratic equation can be found by the formula $\frac{-b}{a}$ where a and b are the coefficients of x^2 and x, respectively. Since $a = 4$ and $b = -17$, the sum of solutions is $-\frac{-17}{4} = \frac{17}{4}$.

Distractor Explanations: Choice A is incorrect. This is the solution to the quadratic equation. **Choice B** is incorrect. This is the value of product of the solutions which can be solved using the formula $\frac{c}{a}$. **Choice C** is incorrect. This is the solution to the quadratic equation.

121. **Level:** Medium | **Skill/Knowledge:** Nonlinear equations in one variable and systems of equations in two variables | **Testing Point:** Finding the roots of a quadratic equations

Key Explanation: 6 is the correct answer. The product of the roots of a quadratic equation can be solved using the formula $\frac{c}{a}$ where a is the coefficient of x^2 and c is the constant. Since $a = 3$ and $c = -6$ the product of the roots is $\frac{-6}{3} = -2$. If one of the roots is $-\frac{1}{3}$, then, $\left(-\frac{1}{3}\right)s = -2$, where s is the second root of the equation. Multiplying both sides of the equation by -3 yields $s = 6$.

122. **Level:** Easy | **Skill/Knowledge:** Nonlinear equations in one variable and systems of equations in two variables | **Testing Point:** Solving a quadratic equation

Key Explanation: 0 is the correct answer. First, determine the factors of the equation to find the solutions. The equation $b^2 - \frac{1}{4} = 0$ features difference of two squares, which indicates that the equation can be factored into conjugates.

Factoring the left side of the equation yields $\left(b + \frac{1}{2}\right)\left(b - \frac{1}{2}\right) = 0$. Equating each factor to 0 yields $b = \frac{1}{2}$ or $b = -\frac{1}{2}$. Adding the two factors yields $\frac{1}{2} + \left(-\frac{1}{2}\right)$ or 0.

123. **Level:** Medium | **Skill/Knowledge:** Nonlinear equations in one variable and systems of equations in two variables | **Testing Point:** Finding the solutions of a quadratic equation

Key Explanation: Choice B is correct. Expanding the left side of the equation yields $3(x^2 - 4x + 4) + 5 = 3x^2 + 12x - 17$.

Using distributive properties yields $3x^2 - 12x + 17 = 3x^2 + 12x - 17$, which is equivalent to $-12x + 17 = 12x - 17$. Adding $12x$ and 17 to both sides of the equation yields $34 = 24x$. Dividing both sides of the equation by 24 yields $\frac{17}{12} = x$ or $x = \frac{17}{12}$. Since there is only one value of x, **Choice B** is the correct answer.

Distractor Explanations: Choice A is incorrect. For an equation to have no solution, simplifying the equation would result in a false statement. **Choice C** is incorrect. Linear systems cannot have two solutions. **Choice D** is incorrect. For a system to have infinite solutions, simplifying the equation would result in a false statement.

124

The Law of Universal Gravitation can be represented by the formula $g=\frac{GM}{R^2}$ where g is the acceleration due to gravity, G is a constant, M is mass, and R is the distance. Which of the following gives the value of R in terms of g, G, and M?

A) $R=\sqrt{\frac{GM}{g}}$

B) $R=\sqrt{\frac{g}{GM}}$

C) $R=\frac{g}{GM}$

D) $R=\frac{gM}{G}$

125

$f(x) = (x - 4)(x + 2)$ and $g(x) = 4x - 17$

For what value of x is the statement $f(x) - g(x) = 0$ true?

A) -3

B) $\sqrt{3}$

C) 3

D) 6

126

What is the value of x when $\frac{1}{x+3}=(x^2-x-12)^{-1}$?

A) -3

B) 3

C) 4

D) 5

127

If $a=\frac{1}{2}$, then what is the value of the expression $\frac{27^a}{3^a}$?

128

Which of the graphs below most closely represents the function $y = 2^x$?

A) 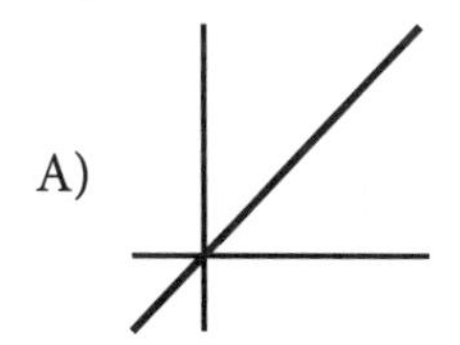

B) 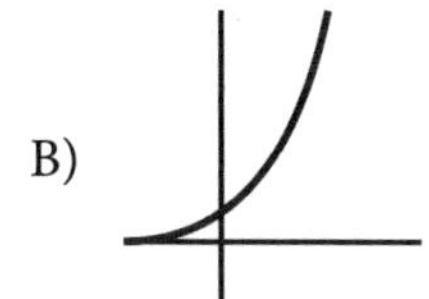

C) 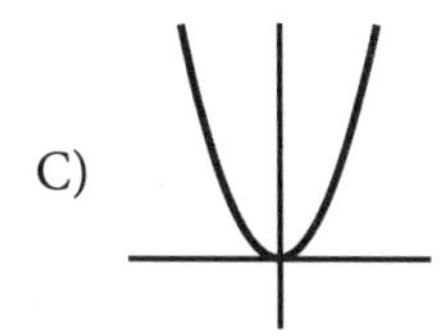

D) 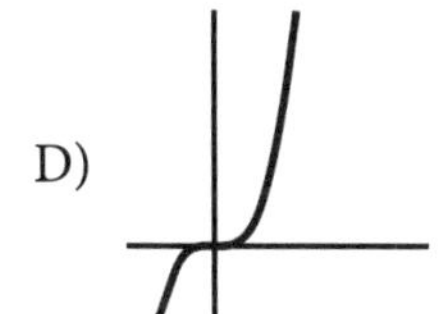

129

The formula for air resistance can be represented by the equation $F = -cv^2$, where F is the force of air resistance, c is the air constant, and v is the velocity of the object. The negative sign indicates that the direction of air resistance is opposite to the direction of motion of the object. Which of the following equations represents v in terms of c and F?

A) $v = cF^2$

B) $v = \sqrt{-\dfrac{c}{4F}}$

C) $v = \sqrt{\dfrac{F}{-c}}$

D) $v = \sqrt{\dfrac{F}{c}}$

130

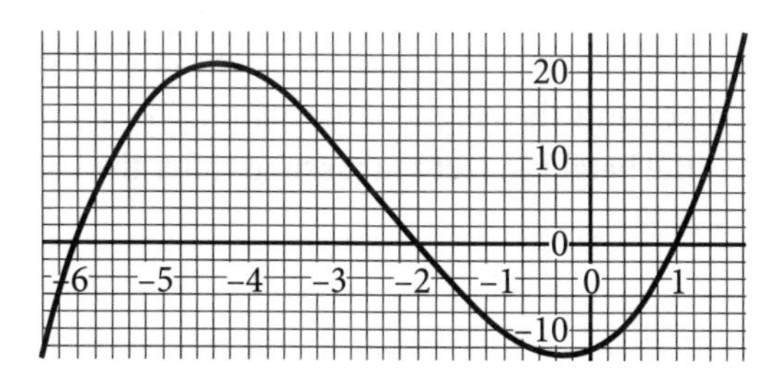

Which equation below represents the graph in the *xy*-plane above?

A) $y = (x^2 + 5x - 6)(x + 2)$

B) $y = (x^2 + 5x - 6)(x + 3)$

C) $y = (x^2 + 3x + 2)(x + 1)$

D) $y = (x^2 + 3x + 2)(x + 2)$

131

A golfer hits a golf ball at a driving range. If the equation $h = -t^2 + 15t$ (t = *time in seconds*) represents the height of the ball (h) in feet, what is the height in feet four seconds after the ball was hit?

132

For the graph of a parabola with a positive lead coefficient, and a vertex at (2, 1), which quadrant(s) does the graph not pass through?

i. Quadrant II

ii. Quadrant III

iii. Quadrant IV

A) i

B) i and ii

C) ii and iii

D) i, ii, and iii

133

If the x-intercepts of the graph of a quadratic function are (–1, 0) and (5, 0), and the graph has not been dilated or reflected from the parent function, what is the distance between the points on the graph intercepted by the line $y = 7$?

A) 3

B) 8

C) 9

D) 10

134

The formula for the break-even point of the manufacturing of a certain product is $b = \frac{f}{s} - c$ where (b) is the break-even point, (f) is the fixed costs, (s) is the sales price, and (c) is the cost to make each unit. Which of the following correctly expresses (s) in terms of b, f, and c?

A) $s = \frac{f}{b+c}$

B) $s = \frac{b+c}{f}$

C) $s = \frac{f}{b} - c$

D) $s = \frac{b+f}{c}$

135

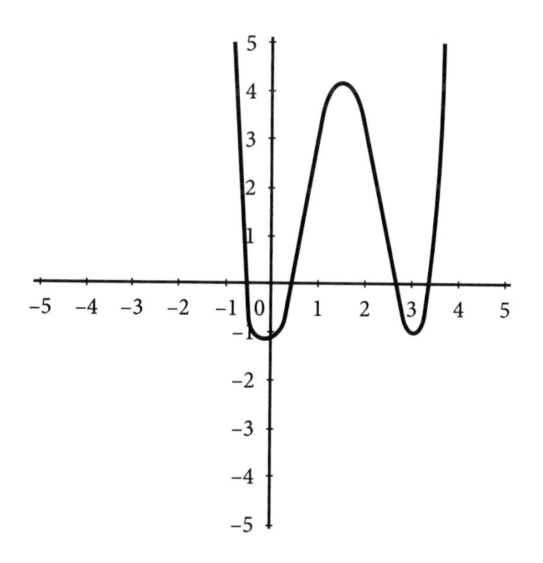

Which of the following could be the equation of the graph above?

A) $x^2(x-3)^2-1$

B) $x(x+3)^2-1$

C) $x^2(x-3)^2$

D) $x^2(x+3)^2+1$

136

In the equation $a+3=\dfrac{1}{a+3}$, which of the following is a possible value for $a+3$?

A) −3

B) 1

C) 2

D) 3

137

Which of the following is a value for b for which the expression $\dfrac{b^2+30b+52}{b^2-169}$ is undefined?

A) −2

B) 0

C) 2

D) 13

138

The range of the function $g(x)$ is the set of real numbers greater than or equal to $-\dfrac{9}{4}$. If the zeros of the function $g(x)$ are −3 and 0, which of the following could be $g(x)$?

A) $(x+5)(x-2)+1$

B) $(x+5)(x-2)+10$

C) $(x+5)(x-2)+5$

D) $(x+5)(x-2)+2$

139

The formula for specific resistance can be expressed as $R = p\frac{1}{A}$ where p is a constant and A is the cross-sectional area. Which of the following gives A in terms of R and p?

A) $A = R\frac{1}{p}$

B) $A = \frac{p}{R}$

C) $A = pR$

D) $A = \frac{1}{pR}$

140

$$f(x) = ax^2 - 22$$

For the function $f(x)$, a is a constant. If $f(-2) = -14$, what is the value of $f(3)$?

A) −4

B) −2

C) 2

D) 4

141

For the function $f(x) = 4x + x^2 - 12$, what is the value of (x) when $f(x)$ is at its minimum?

142

If $6 \times 6^{(s-1)} = 1{,}296$, what is the value of s?

A) 1

B) 2

C) 3

D) 4

143

Which of the following graphs below best represents the equation $y = 3^x - 2$?

A)

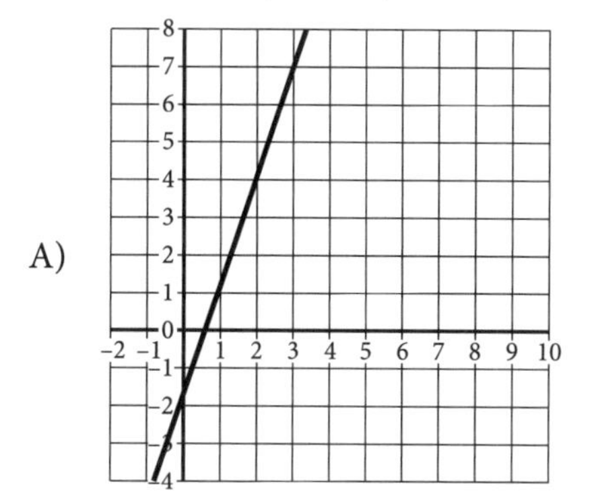

B)

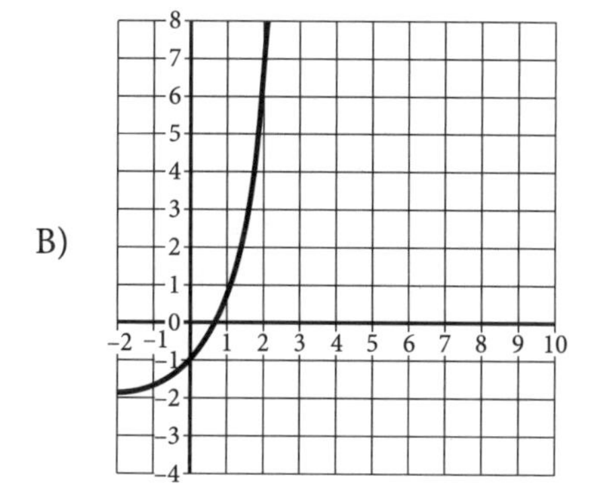

C)

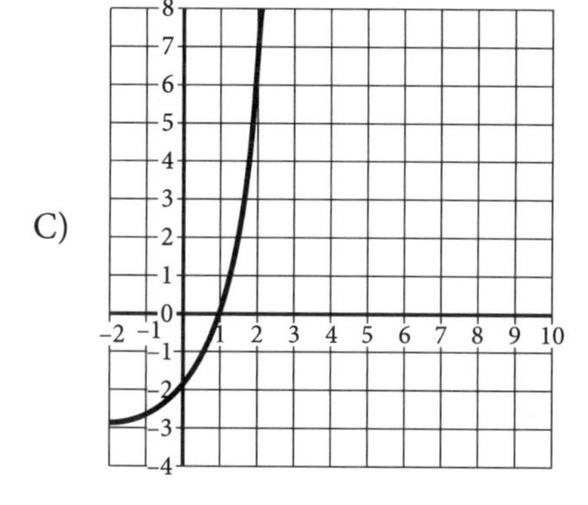

D)

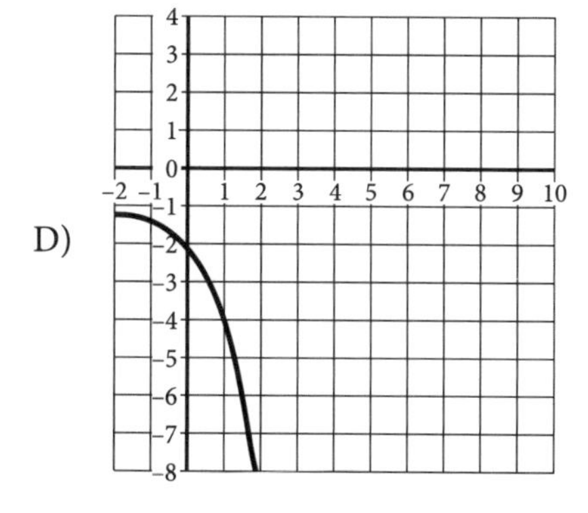

144

If $h(n) = n^3 - 8$, what is the value of n when $h(n) = 19$?

A) –8

B) –2

C) 3

D) 9

145

The population of a town is represented by $p(t) = 3{,}402(1.012)^t$, where t is the number of years. By what percentage does the population of the town grow every t years?

A) 1.2%

B) 12%

C) 101.2%

D) 3,402%

146

Which of the following shows an equation of a parabola in vertex form?

A) $y = 3x^2 + 6x + 14$

B) $y = 3(x^2 + 55)$

C) $y = 3x^2 + 55$

D) $y = 3(x + 2)^2 + 13$

147

What is the positive value of x in the equation below?

$$|x+4|=8$$

148

If $g(x) = 2x^2 - 25x + 20$ is a function in x, for what value of x, $g(x)$ reaches its minimum value?

A) $\frac{25 \pm \sqrt{465}}{4}$

B) $\frac{25}{4}$

C) 20

D) 25

149

$$\sqrt{g+3} = f\sqrt{5}$$

If $f = 2$, what is the value of g?

150

If $|x+3| = \frac{11}{5}$, what is the sum of all values of x?

151

If $f(x)=\frac{3}{2}x+2$ and $f(x)=3$, what is the value of $(x)^{-1}$?

A) $\frac{2}{3}$

B) $\frac{3}{2}$

C) 2

D) 3

152

What is the value of $\left(x+\sqrt{3}\right)^2+3$ when $x = 0$?

A) −3

B) 0

C) 3

D) 6

Nonlinear functions

153

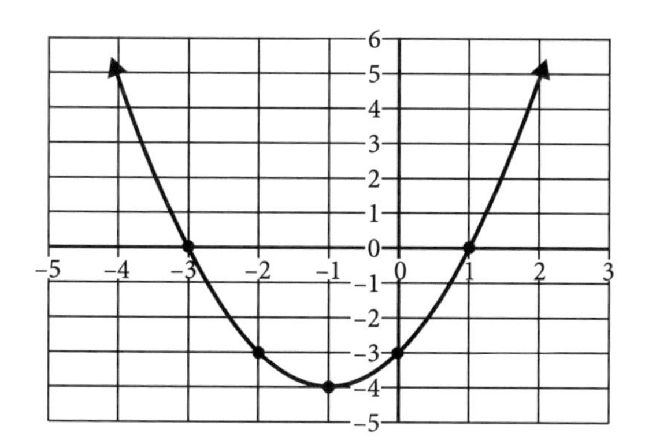

Which of the following equations, when graphed on the (x, y) plane above, expresses the following function in vertex form?

A) $y=\frac{1}{2}x-\frac{5}{2}$

B) $y=(x+1)^2-4$

C) $y=(x-1)(x+3)$

D) $y=(x+1)^2+4$

154

What is the value of $-x$ if $x > 0$ and $f(x) = \left(\frac{x^2+x-12}{(x+4)}\right)^2$ when $f(x) = 25$?

155

How many real solutions does $y = 2x^2 - 3x + 5$ have?

156

If $f(x) = x^2 + x$, which of the following expressions is equal to $f(x + 2)$?

A) $(x + 4)(x + 1)$

B) $(x + 2)(x + 3)$

C) $x^2 + 5x - 6$

D) $x^2 + 2x + 2$

157

$$f(x) = x^2 - 36$$
$$g(x) = x^3 + 18x^2 + 108x + 216$$

Which of the following is equivalent to $\frac{g(x)}{f(x)}$?

A) $\frac{x^2+12x+36}{x-6}$

B) $\frac{1}{x-6}$

C) $\frac{1}{x+6}$

D) $(x + 6)^3$

158

Every year, Ron adds twice the number of model airplanes to his collection from the year before. Which of the following type of function best describes the number of model airplanes Ron has in his collection with respect to time?

A) Increasing exponential

B) Decreasing exponential

C) Increasing linear

D) Decreasing linear

159

If a parabola represented by the equation $f(x) = -2(x+2)^2 + 12$ is translated to $g(x) = -2x^2 + 4x + 14$, which of the following statements best describes the horizontal transformation?

A) The function moves 3 units to the right.

B) The function moves 3 units to the left.

C) The function moves 4 units left.

D) The function moves 4 units right.

160

How many integer solutions does the equation below have?

$$|x+3| < 4$$

A) −6

B) 0

C) 2

D) 7

161

A substance has a mass of 312 g and a half-life of 120 days. Its half-life is given by the equation $p(d) = 312\left(\frac{1}{2}\right)^{\frac{d}{120}}$, where d is time in days. If the equation is rewritten using m months as a time interval, which of the following best represent the equation of the half-life? (Assume that every month has 30 days)

A) $p(m) = 312\left(\frac{1}{2}\right)^{\frac{m}{120}}$

B) $p(m) = 312\left(\frac{1}{2}\right)^{30m}$

C) $p(m) = 312\left(\frac{1}{2}\right)^{\frac{m}{30}}$

D) $p(m) = 312\left(\frac{1}{2}\right)^{\frac{m}{4}}$

162

What is the y-intercept of the exponential equation $y = -(6)^x - 1$?

163

A phone company creates a model that shows the number of phones they have sold in the past 10 years. The number of phones sold in the first year was 1,600,000. The number of phones then increased by 72% every year after that. Which of the following best represents the model equation of the number of phones sold by the company where t is the number of years?

A) $p(t) = 1{,}600{,}000(0.72)^t$

B) $p(t) = 1{,}600{,}000(1.72)^t$

C) $p(t) = 1{,}600{,}000(0.28)^t$

D) $p(t) = 1{,}600{,}000(1.72)^{12t}$

164

The revenue of a consulting firm is described by the model $f(x)$ shown below, where x is the number of years and $0 \le x \le 9$. Which of the following best describes $f(2) = 40{,}140.8$?

$$f(x) = 32{,}000(1.12)^x$$

A) The revenue increases by $40,140.8 after 2 years.

B) The revenue of the consulting firm was $40,140.8 at the beginning of year 2.

C) The revenue of the consulting firm is $40,140.8 at the end of year 2.

D) The revenue of the consulting firm decreases to $40,140.8 after 2 years.

165

Chemical w was observed and the researcher came up with the model $p(d) = 300(3)^{\frac{d}{5}}$, where $p(d)$ is the mass of chemical w after d days.

Which statement would best describe $(3)^{\frac{d}{5}}$?

A) The mass of chemical w triples every 5 days.

B) The mass of chemical w triples every d days.

C) The mass of chemical w is five times its initial mass after 3 days.

D) The initial mass of chemical w.

166

A botanist observes the growth of a tree for 25 years and comes up with the model in meters. The function is $h(t) = 2.3(1.12)^t$, where t is the number of years. Which of the following statements is NOT true about the model?

A) The initial height of the tree is 230 cm.

B) The height of the tree increases by 12% annually.

C) The height of the tree after the first year of observation is 2.576 m.

D) The height of the tree increases by 112% annually.

124. **Level:** Medium | **Skill/Knowledge:** Nonlinear functions | **Testing Point:** Solving for one variable in terms of another

Key Explanation: Choice A is correct. Multiplying both sides of the equation by R^2 yields $R^2g = GM$. Dividing both sides of the equation by g yields $R^2 = \frac{GM}{g}$. Getting the square root of both sides of the equation yields $R = \sqrt{\frac{GM}{g}}$.

Distractor Explanations: Choice B is incorrect and is the result of incorrectly multiplying and dividing variables. **Choice C** is incorrect and is the result of incorrectly multiplying and dividing variables. **Choice D** is incorrect and is the result of incorrectly multiplying and dividing variables.

125. **Level:** Medium | **Skill/Knowledge:** Nonlinear functions | **Testing Point:** Performing operations with functions

Key Explanation: Choice C is correct. To find the value of x that makes $f(x) - g(x) = 0$, plug each equation of $f(x)$ and $g(x)$. This yields $(x - 4)(x + 2) - (4x - 17) = 0$. Using the distributive property yields $x^2 - 2x - 8 - 4x + 17 = 0$. Combining like terms yields $x^2 - 6x + 9 = 0$. Factoring the perfect square trinomial yields $(x - 3)^2 = 0$. Equating the factor to zero yields $x - 3 = 0$. Therefore, $x = 3$.

Distractor Explanations: Choice A is incorrect and may result from not equating the factor to zero in solving for x. **Choice B** is incorrect and may result from a conceptual or calculation error. **Choice D** is incorrect and may result from adding the values of x from the two factors.

126. **Level:** Hard | **Skill/Knowledge:** Nonlinear functions | **Testing Point:** Solving rational equations

Key Explanation: Choice D is correct. Making the right side of the equation have positive exponent yields $\frac{1}{x+3} = \frac{1}{x^2 - x - 12}$. Cross multiplying to eliminate the fractions yields $x^2 - x - 12 = x + 3$. Factoring the left side of the equation yields $(x - 4)(x + 3) = (x + 3)$. Dividing both sides of the equation by $(x + 3)$ yields $x - 4 = 1$. Adding 4 to both sides of the equation yields $x - 4 + 4 = 1 + 4$. Therefore, $x = 5$.

Distractor Explanations: Choice A is incorrect and may result in the denominator of the left side being undefined. **Choice B** is incorrect and may result from multiplying -1 by the trinomial. **Choice C** is incorrect and may result from solving x from the equation $x - 4 = 0$.

127. **Level:** Hard | **Skill/Knowledge:** Nonlinear functions | **Testing Point:** Evaluating equations with exponents

Key Explanation: 3 is the correct answer. To evaluate the given expression, make the base the same. Converting 27^a yields $(3^3)^a$ or 3^{3a}. Simplify the expression $\frac{3^{3a}}{3^a}$ by subtracting the exponents which yield 3^{3a-a} or 3^{2a}. Substituting the value of a yields $3^{2(½)}$ or 3^1. Therefore, the expression is equal to 3.

128. **Level:** Easy | **Skill/Knowledge:** Nonlinear functions | **Testing Point:** Determining a graph of an exponential function given its equation

Key Explanation: Choice B is correct. The equation $y = 2^x$ is an exponential function, and the graph in **Choice B** is the only graph that features exponential growth.

Distractor Explanations: Choice A is incorrect because it features a graph of a linear function. **Choice C** is incorrect because it features a graph of a quadratic function. **Choice D** is incorrect because it features a graph of a cubic function.

129. **Level:** Easy | **Skill/Knowledge:** Nonlinear functions | **Testing Point:** Solving for one variable in terms of another

Key Explanation: Choice C is correct. In order to rewrite the equation in terms of v, one must isolate that variable on one side of the equation:

$$F = -cv^2$$

$$v^2 = \frac{F}{-c}$$

$$v = \sqrt{\frac{F}{-c}}.$$

Distractor Explanations: Choice A is incorrect because when they are rearranged to isolate F, they do not equal to $F = -cv^2$. **Choice B** is incorrect because when they are rearranged to isolate F, they do not equal to $F = -cv^2$. **Choice D** is incorrect because when they are rearranged to isolate F, they do not equal to $F = -cv^2$.

130. **Level:** Easy | **Skill/Knowledge:** Nonlinear functions | **Testing Point:** Making connections between algebraic equation representations and a graph

Key Explanation: Choice A is correct. To determine the equation of the polynomial in the graph, recognize that all answers are written in some version of intercept form. Then identify the x-intercepts of the graph, which lie at (–6, 0), (–2, 0), and (1, 0). The equation of the graph can then be written as $y = (x + 6)(x - 1)(x + 2)$. This answer can be simplified as $y = (x^2 + 5x - 6)(x + 2)$ by multiplying the first two factors together.

Distractor Explanations: Choice B is incorrect and may result from using wrong x-intercepts. **Choice C** is incorrect and may result from using wrong x-intercepts. **Choice D** is incorrect and may result from using wrong x-intercepts.

131. **Level:** Easy | **Skill/Knowledge:** Nonlinear functions | **Testing Point:** Using a quadratic function

Key Explanation: 44 is correct. The height of the ball after (t) seconds can be found by substituting 4 seconds for (t) in the equation:

$h = -(4)^2 + 15(4)$

$h = -16 + 60$

$h = 44\ ft.$

132. **Level:** Easy | **Skill/Knowledge:** Nonlinear functions | **Testing Point:** Creating a quadratic and interpreting its location on a graph

Key Explanation: Choice C is correct. A parabola with a positive lead coefficient opens upward. Since the vertex of the given parabola is at (2, 1) then the lowest value of the graph is at $y = 1$. This means that the graph does not go below $y = 1$. Therefore, the parabola does not pass through the quadrants below $y = 0$ or the x-axis which are Quadrants III and IV.

Distractor Explanations: Choice A is incorrect and may result from an error in interpreting the data in the equation of a parabola. **Choice B** is incorrect and may result from an error in interpreting the data in the equation of a parabola. **Choice D** is incorrect and may result from an error in interpreting the data in the equation of a parabola.

133. **Level:** Hard | **Skill/Knowledge:** Nonlinear functions | **Testing Point:** Creating quadratic equation and using transformations

Key Explanation: Choice B is correct. To find where the function intercepts the function $y =$

7, first write the quadratic equation using the given x-intercepts. Using the intercept form, the equation becomes $y = (x + 1)(x - 5)$. Substituting the value of y yields $7 = (x + 1)(x - 5)$. Using the distributive method on the right side of the equation yields $7 = x^2 - 4x - 5$. Subtracting 7 from both sides yields $0 = x^2 - 4x - 12$. Factoring the equation to find the values of x when $y = 7$ yields $0 = (x + 2)(x - 6)$. Equating both factors to zero yields $x = -2$ or $x = 6$. The two points that intercept the other function are then $(-2, 7)$ and $(6, 7)$, so the distance between the two points is $6 - (-2) = 8$.

Distractor Explanations: Choice A is incorrect and reflects errors in finding solutions and intersections between two functions. **Choice C** is incorrect and reflects errors in finding solutions and intersections between two functions. **Choice D** is incorrect and reflects errors in finding solutions and intersections between two functions.

134. **Level:** Easy | **Skill/Knowledge:** Nonlinear functions | **Testing Point:** Solving for one variable in an equation

Key Explanation: Choice A is correct. To solve for s, add c to both sides:

$$b = \frac{f}{s} - c$$

which yields $b + c = \frac{f}{s}$.

Then multiply both sides by s which yields $(b + c)s = f$.

Then divide both sides of the equation by $b + c$:

which yields $s = \frac{f}{b+c}$.

This gives **Choice A** as the correct answer.

Distractor Explanations: Choice B is incorrect and is the result of errors in rearranging equations with multiple variables. **Choice C** is incorrect and is the result of errors in rearranging equations with multiple variables. **Choice D** is incorrect and is the result of errors in rearranging equations with multiple variables.

135. **Level:** Medium | **Skill/Knowledge:** Nonlinear functions | **Testing Point:** Making connections between graphs and their algebraic representations

Key Explanation: Choice A is correct. The graph crosses the x-axis in 4 places and thus has 4 solutions. It is then a quartic equation of highest power x^4. This eliminates **Choice B** as when this expression is multiplied out it yields a highest power of 3. The graph crosses the x-axis at approximately -0.5, 0.5, 2.5, and 3.5 and thus has zeros at these points. This eliminates **Choice C** as it has zeros at $x = 0$ and $x = 3$. The function $y = x^2(x - 3)^2 - 1$, or **Choice A**, has a vertical shift down 1 from the graph of $x^2(x - 3)^2$. That would move the graph 1 unit below the x-axis from the parent function whose lowest point would be at the x-axis. This matches the graph. **Choice D** would move the graph one unit up which is not represented in the diagram.

Distractor Explanations: Choice B is incorrect and reflects error in interpreting the graph as stated above. **Choice C** is incorrect and reflects error in interpreting the graph as stated above. **Choice D** is incorrect and reflects error in interpreting the graph as stated above.

136. **Level:** Easy | **Skill/Knowledge:** Nonlinear functions | **Testing Point:** Solving a quadratic equation by square root method

Key Explanation: The first step in solving the equation $a + 3 = \frac{1}{a+3}$ is through cross

multiplication. This yields $(a + 3)^2 = 1$. Taking the square root of both sides of the equation results in $a + 3 = \pm 1$.
Only 1 is in the answer choices. Thus, the correct answer is **Choice B**.

Distractor Explanations: Choice A is incorrect and is the result of arithmetic error. **Choice C** is incorrect and is the result of arithmetic error. **Choice D** is incorrect and is the result of arithmetic error.

137. **Level:** Medium | **Skill/Knowledge:** Nonlinear functions | **Testing Point:** Using difference of two squares and knowledge of rational expressions

Key Explanation: Choice D is correct. A rational expression is undefined where its denominator is equal to zero. Thus, in this problem where $b^2 - 169 = 0$. Using the difference of two squares $b^2 - 169$ is equivalent to $(b - 13)(b + 13)$. Setting $(b - 13)$ and $(b + 13)$ each equal to zero results in $b = \pm 13$. Only $b = 13$ is an answer choice. Thus, **Choice D** is the correct answer.

Distractor Explanations: Choice A is incorrect and is the result of arithmetic error in setting the denominator equal to zero. **Choice B** is incorrect and is the result of arithmetic error in setting the denominator equal to zero. **Choice C** is incorrect and is the result of arithmetic error in setting the denominator equal to zero.

138. **Level:** Hard | **Skill/Knowledge:** Nonlinear functions | **Testing Point:** Determining the function from the zeros and the range

Key Explanation: Choice B is correct. Test the answer choices by expanding out each quadratic function in the answers to see which yields zeros of -3 and 0. Expanding out **Choice B** yields $(x + 5)(x - 2)$ which expands to $x^2 + 3x - 10$. Adding the 10 to this expanded polynomial results in $x^2 + 3x$. Factoring out an x results in $x(x + 3)$ and setting equal to zero yields the solution, or zeros, of $x = 0$ and $x = -3$, or **Choice B**. This answer choice also results in a vertex of the parabola of $\left(\frac{-3}{2}, \frac{-9}{4}\right)$. Since the parabola opens up, its minimum value is $\frac{-9}{4}$ which makes the range of $g(x)$ greater than or equal to $\frac{-9}{4}$.

Distractor Explanations: Choice A is incorrect and is the result of errors in factoring polynomials to find solutions or knowledge of parabolas and their equation forms. **Choice C** is incorrect and is the result of errors in factoring polynomials to find solutions or knowledge of parabolas and their equation forms. **Choice D** is incorrect and is the result of errors in factoring polynomials to find solutions or knowledge of parabolas and their equation forms.

139. **Level:** Easy | **Skill/Knowledge:** Nonlinear functions | **Testing Point:** Solving for one variable in terms of another

Key Explanation: Choice B is correct. In order to express the equation in terms of A, isolate variable A on the left side of the equation.

Multiplying both sides of the equation by A yields $RA = p$.

Dividing both sides of the equation by R yields $A = \frac{p}{R}$.
Therefore, **Choice B** is the correct answer.

Distractor Explanations: Choice A is incorrect as when it is rearranged to isolate R it is not equal to $R = p\frac{1}{A}$. **Choice C** is incorrect as when it is rearranged to isolate R it is not equal to $R = p\frac{1}{A}$. **Choice D** is incorrect as when it is rearranged to isolate R it is not equal to $R = p\frac{1}{A}$.

140. **Level:** Easy | **Skill/Knowledge:** Nonlinear functions | **Testing Point:** Solving for a variable in a quadratic function

Key Explanation: Choice A is correct. Start by determining the value for a by replacing x with -2 and $f(x)$ with -14.

This yields $-14 = a(-2)^2 - 22$.

Simplifying and adding 22 to both sides of the equation yields $8 = a(4)$.

Dividing both sides of the equation by 4 yields $2 = a$ or $a = 2$.

Replace a with 2 and x with 3 to find the value of $f(3)$ which yields

$f(3) = 2(3)^2 - 22 = 2 \times 9 - 22 = 18 - 22 = -4$.

Distractor Explanations: Choice B is incorrect and reflects error in arithmetic or solving algebraic equations. **Choice C** is the value of a and reflects a misunderstanding of the question. **Choice D** is incorrect and reflects error in arithmetic or solving algebraic equations.

141. **Level:** Hard | **Skill/Knowledge:** Nonlinear functions | **Testing Point:** Working with different forms of parabola equation and vertex

Key Explanation: -2 is the correct answer. When graphed, the vertex of the parabola is clearly the minimum $f(x)$ value. To determine the vertex, rewrite the equation in vertex form $f(x) = (x - h)^2 + k$ where (h, k) is the vertex. Completing the squares and grouping the perfect square trinomial yields $f(x) = (x^2 + 4x + 4) - 16$. Simplifying the equation yields $f(x) = (x + 2)^2 - 16$. Hence, the vertex is $(-2, -16)$. Therefore, $x = -2$ when $f(x)$ is at its minimum.

142. **Level:** Easy | **Skill/Knowledge:** Nonlinear functions | **Testing Point:** Using exponent rules to solve nonlinear function

Key Explanation: Choice D is correct. Using the exponent rules for multiplication, the left side of the equation can be simplified by adding the exponents. This yields $6 \times 6^{s-1} = 6^{1+s-1} = 6^s = 1{,}296$. It follows that $s = 4$ because $6^4 = 1{,}296$.

Distractor Explanations: Choice A is incorrect and is the result of errors in solving equation and exponent. **Choice B** is incorrect and is the result of errors in solving equation and exponent. **Choice C** is incorrect and is the result of errors in solving equation and exponent.

143. **Level:** Hard | **Skill/Knowledge:** Nonlinear functions | **Testing Point:** Matching a graph to its exponential equation

Key Explanation: Choice B is correct. The equation $y = 3^x - 2$, is an exponential growth function. Hence, the possible answer is either **Choice B** or C. To determine which one is correct, find the y-intercept of the graph by substituting 0 with x. This yields $y = 3^0 - 2$ or $y = -1$. Since the graph in **Choice B** intersects the y-axis at $(0, -1)$, it is the correct answer.

Distractor Explanations: Choice A is incorrect. The graph of **Choice A** is linear. **Choice C** is incorrect. The y-intercept of the graph is -2. **Choice D** is incorrect. This is a decreasing exponential graph; however, the given equation requires an increasing exponential graph.

144. **Level:** Easy | **Skill/Knowledge:** Nonlinear functions | **Testing Point:** Solving a cubic function

Key Explanation: Choice C is correct. To find the value of n, equate the equations to each other which yields $19 = n^3 - 8$. Adding 8 to both sides of the equation yields $27 = n^3$ or $n^3 = 27$. Getting the cube root of both sides of the equation yields $n = 3$.

Distractor Explanations: Choice A is incorrect. This is the y-intercept of the function $h(n)$. **Choice B** is incorrect. This is the cube root of the y-intercept. **Choice D** is incorrect. This is the value of n^3.

145. **Level:** Medium | **Skill/Knowledge:** Nonlinear functions | **Testing Point:** Finding the growth rate and growth factor

Key Explanation: Choice A is correct. The growth factor is 1.012 from the given equation. The growth factor is increasing since it is greater than 1. To find the growth rate, calculate the value of r in $1 + r = 1.012$. Subtracting 1 from both sides of the equation yields $r = 0.012$ or $r =$ 1.2% in percentage. This would be the percentage increase in the population of the town every t years.

Distractor Explanations: Choice B is incorrect and may result from an error in converting the decimal to fraction. **Choice C** is incorrect. This is the growth factor in percentage form. **Choice D** is incorrect. The value 3,402 is the initial population.

146. **Level:** Easy | **Skill/Knowledge:** Nonlinear functions | **Testing Point:** Recognizing the vertex form of a parabola

Key Explanation: Choice D is correct. The equation of a parabola in vertex form is $y = a(x - h)^2 + k$, where (h, k) is the vertex. **Choice D** is in the vertex form. The vertex of this line $y = 3(x + 2)^2 + 13$ is $(-2, 13)$.

Distractor Explanations: Choice A is incorrect. This equation of a parabola is written in standard form. **Choice B** is incorrect. This equation of a parabola is written in standard form. **Choice C** is incorrect. This equation of a parabola is written in standard form.

147. **Level:** Easy | **Skill/Knowledge:** Nonlinear functions | **Testing Point:** Solving absolute value equations

Key Explanation: To solve for the positive value of x, equate the contents of the absolute value to the positive value of the constant. This yields $x + 4 = 8$. Subtracting 4 from both sides of the equation yields $x = 4$.

148. **Level:** Hard | **Skill/Knowledge:** Nonlinear functions | **Testing Point:** Solving quadratic equations

Key Explanation: Choice B is correct. To find the minimum value of x in a quadratic equation, use the formula $x_{min} = -\frac{b}{2a}$. Comparing the given equation $g(x) = 2x^2 - 25x + 20$ to the general form equation $ax^2 + bx + c = 0$ yields $a = 2$ and $b = -25$. Substituting the values of a and b yields $x_{min} = -\frac{(-25)}{2(2)}$. Therefore, $x_{min} = \frac{25}{4}$.

Distractor Explanations: Choice A is incorrect and may result from getting the value of x when $g = 0$. **Choice C** is incorrect and may result from getting the value of g when $x = 0$. **Choice D** is incorrect and may result from a conceptual or calculation error.

149. **Level:** Easy | **Skill/Knowledge:** Nonlinear functions | **Testing Point:** Solving square root functions

Key Explanation: The correct answer is 17. Substituting $f = 2$ to the equation yields

$\sqrt{g+3}=(2)\sqrt{5}$. Squaring both sides of the equation to eliminate the radicals yields $g + 3 = 4(5)$. Subtracting 3 from both sides of the equation yields $g = 4(5) - 3$ or $g = 17$.

150. **Level:** Medium | **Skill/Knowledge:** Nonlinear functions | **Testing Point:** Solving an absolute value equation

Key Explanation: The absolute value indicates that there are two solutions to the equation, which can be found by creating two equations as follows $x+3=\frac{11}{5}$ and $x+3=-\frac{11}{5}$. By subtracting 3 from both sides of the two equations yields $x=\frac{11}{5}-3$ and $x=-\frac{11}{5}-3$. Simplifying the two equations yields $x=-\frac{4}{5}$ and $x=-\frac{26}{5}$. Adding the two values of x yields $-\frac{4}{5}+\left(-\frac{26}{5}\right)$ or $-\frac{30}{5}$.
Therefore, the sum of all the values of x is -6.

151. **Level:** Hard | **Skill/Knowledge:** Nonlinear functions | **Testing Point:** Evaluating a linear function for a value and then determining the value of a rational function

Key Explanation: Choice B is correct. First, use the information provided to determine the value of x, before finding the value of $(x)^{-1}$. Combining the two equations yields $\frac{3}{2}x+2=3$. Subtracting 2 from both sides of the equation yields $\frac{3}{2}x=1$. Multiplying 2 and dividing 3 from both sides of the equation yields $x=\frac{2}{3}$. Hence, $(x)^{-1}=\frac{1}{x}=\frac{3}{2}$.

Distractor Explanations: Choice A is incorrect because it is the value of x. **Choice C** is incorrect and may result from a conceptual or calculation error. **Choice D** is incorrect and may result from a conceptual or calculation error.

152. **Level:** Easy | **Skill/Knowledge:** Nonlinear functions | **Testing Point:** Evaluating a quadratic expression given an input value

Key Explanation: Choice D is correct. Begin by substituting $x = 0$ into the expression $\left(x+\sqrt{3}\right)^2+3$ which yields $\left(0+\sqrt{3}\right)^2+3$. Simplifying the expression yields $\left(\sqrt{3}\right)^2+3$ or $3 + 3$. Therefore, the value of the expression is 6.

Distractor Explanations: Choice A is incorrect and may result from a conceptual or calculation error. **Choice B** is incorrect and may result from a conceptual or calculation error. **Choice C** is incorrect and may result from a conceptual or calculation error.

153. **Level:** Easy | **Skill/Knowledge:** Nonlinear functions | **Testing Point:** Creating quadratic equation from its graph

Key Explanation: Choice B is correct. The graph shown represents a quadratic function. To create the quadratic equation, use the vertex form $y - k = (x - h)^2$ where (h, k) is the vertex of the graph. Therefore, when $(-1, -4)$ is substituted into the vertex form equation, it yields $y + 4 = (x + 1)^2$. This is equivalent to $y = (x + 1)^2 - 4$.

Distractor Explanations: Choice A is incorrect because it represents a linear equation, not a quadratic equation. **Choice C** is incorrect as it is in intercept form, not vertex form. **Choice D** is incorrect because it includes the opposite sign for the y-value of the vertex.

154. **Level:** Hard | **Skill/Knowledge:** Nonlinear functions | **Testing Point:** Solving a rational equation

Key Explanation: –8 is the correct answer. Substituting the value of $f(x)$ to equation $f(x) = \left(\frac{x^2+x-12}{(x+4)}\right)^2$ yields $25 = \left(\frac{x^2+x-12}{(x+4)}\right)^2$.

Getting the square root of both sides of the equation yields $\sqrt{25} = \sqrt{\left(\frac{x^2+x-12}{(x+4)}\right)^2}$ or $\pm 5 = \frac{x^2+x-12}{(x+4)}$. Factoring the right side of the equation yields $\pm 5 = \frac{(x+4)(x-3)}{x+4}$ or $\pm 5 = x-3$.

Equating the right side of the equation to both values of the left side yields $5 = x - 3$ or $-5 = x - 3$. Solving for x in both equations yields $x = 8$ or $x = -2$. Since $x > 0$, then $x = 8$. Therefore, the value of $-x$ is -8.

155. **Level:** Easy | **Skill/Knowledge:** Nonlinear functions | **Testing Point:** Using the discriminant of a quadratic equation to find the number and type of solutions

Key Explanation: The answer is 0. To find the number of solutions a function has, use the discriminant formula $b^2 - 4ac$. In this function $a = 2$, $b = -3$, $c = 5$. The discriminant therefore is $(-3)^2 - 4(2)(5) = 9 - 40 = -31$.

If the discriminant is a negative number, this indicates that the function has no real solutions.

156. **Level:** Easy | **Skill/Knowledge:** Nonlinear functions | **Testing Point:** Evaluating a nonlinear function given an algebraic binomial input value

Key Explanation: Choice B is correct. Begin by substituting $(x + 2)$ for x in the expression which yields $(x + 2)^2 + (x + 2)$.

Using distributive property to simplify the expression yields $x^2 + 4x + 4 + x + 2$.

Combining like terms yields $x^2 + 5x + 6$. Factoring the expression yields $(x + 2)(x + 3)$.

Distractor Explanations: Choice A is incorrect and may result from only substituting $(x + 2)$ to x^2. **Choice C** is incorrect and may result from an error in simplifying the expression. **Choice D** is incorrect and may result from adding $x^2 + x$ to $x + 2$.

157. **Level:** Hard | **Skill/Knowledge:** Nonlinear functions | **Testing Point:** Simplifying rational expressions

Key Explanation: Choice A is correct. First substitute the algebraic expressions for $g(x)$ in the numerator and for $f(x)$ in the denominator of the fraction. Factor both numerator and denominator as follows:

The denominator can be factoring using the difference of two squares $a^2 - b^2 =$

$(a + b)(a - b)$ where $a = x$ and $b = 6$ resulting in $f(x) = (x + 6)(x - 6)$.

The numerator is difficult to factor as it can't be factored by grouping. The easiest way to factor it is to assume that $(x + 6)$ might be a factor of it since it is a factor of the denominator. Since all terms in the numerator are positive, $(x - 6)$ cannot be a factor of the numerator. Using long division or synthetic division, divide $(x + 6)$ into the numerator. This results in $x^2 + 12x + 36$. This expression is easily factored into $(x + 6)(x + 6)$.

Cancelling out like terms between the numerator and denominator results in

$$= \frac{(x+6)^3}{x^2-36} = \frac{(x+6)^3}{(x+6)(x-6)} = \frac{(x+6)^2}{x-6} =$$

$\frac{x^2+12x+36}{x-6}$ or **Choice A**.

Distractor Explanations: Choice B is incorrect and is due to errors in factoring or simplifying complex functions. **Choice C** is incorrect and is due to errors in factoring or simplifying complex functions. **Choice D** is incorrect and is due to errors in factoring or simplifying complex functions.

158. **Level:** Easy | **Skill/Knowledge:** Nonlinear functions | **Testing Point:** Recognizing nonlinear equations

Key Explanation: Choice A is correct. Ron is adding planes to his collection at a doubling rate each year. Since his collection does not increase at a constant rate each year, this is considered an exponential increase.

Distractor Explanations: Choice B is incorrect and is the result of error in interpreting functions. **Choice C** is incorrect and is the result of error in interpreting functions. **Choice D** is incorrect and is the result of error in interpreting functions.

159. **Level:** Medium | **Skill/Knowledge:** Nonlinear functions | **Testing Point:** Translating quadratic equations

Key Explanation: Choice A is correct. The vertex of $f(x)$ is $(-2, 12)$. The equation of $g(x) = -2x^2 + 4x + 14$ in vertex form would be $-2(x - 1)^2 + 16$. Hence, the vertex would be $(1, 16)$. The x coordinate translation from -2 to 1 is 3 units to the right.

Distractor Explanations: Choice B is incorrect and may result from an error in interpreting the direction of translation. **Choice C** is incorrect and represents vertical transformation. **Choice D** is incorrect and represents vertical transformation.

160. **Level:** Hard | **Skill/Knowledge:** Nonlinear functions | **Testing Point:** Finding the solutions of an absolute value inequality

Key Explanation: Choice D is correct. To solve for the absolute value linearly, use two inequalities for the positive and negative values.

$+(x+3) < 4$	$-(x+3) < 4$
$x + 3 < 4$	$-x - 3 < 4$
$x < 1$	$-x < 7$
	$x > -7$

Thus, $-7 < x < 1$. Since the value of x is between -7 and 1, the solutions to the above inequality are $0, -1, -2, -3, -4, -5, -6$. Therefore, there are 7 solutions to the inequality above.

Distractor Explanations: Choice A is incorrect and may result from a conceptual or calculation error. **Choice B** is incorrect and may result from a conceptual or calculation error. **Choice C** is incorrect and may result from a conceptual or calculation error.

161. **Level:** Hard | **Skill/Knowledge:** Nonlinear functions | **Testing Point:** Changing units of time in an equation

Key Explanation: Choice D is correct. Every four months or 120 days the mass of the substance is halved. **Choice D** is the only equation that would result in a half-life of 4 months.

Distractor Explanations: Choice A is incorrect. This option suggests that the substance has a half-life of 120 months. **Choice B** is incorrect. This option suggests that the substance has a half-life of 1 day or $\frac{1}{30}$ months.
Choice C is incorrect. This option suggests that the substance has a half-life of 30 months.

162. **Level:** Medium | **Skill/Knowledge:** Nonlinear functions | **Testing Point:** Finding the y-intercept

Key Explanation: –2 is the correct answer. The y-intercept of a line is given when $x = 0$. Substituting 0 for x in the given equation yields $y = -(6)^0 - 1$. Simplifying the equation yields –2.

163. **Level:** Easy | **Skill/Knowledge:** Nonlinear functions | **Testing Point:** Creating an exponential function from data

Key Explanation: Choice B is correct. the equation is an exponential growth function. The equation is $y = a(1 + r)^x$ where a is the initial amount, r is the growth rate and x is the time interval. Since the model is increasing by 72% then $r = 72\%$. The growth factor can be written as $(1 + r) = (1 + 72\%) = 1.72$. Since the initial number of sales is 1,600,000 then $a = 1{,}600{,}000$. Since the time interval is in years, then $x = t$. Therefore, the equation of the model is $p(t)=1{,}600{,}000(1.72)^t$.

Distractor Explanations: Choice A is incorrect. This model represents a function that is decreasing by 28%. **Choice C** is incorrect. This model represents a function that is decreasing by 72%. **Choice D** is incorrect. The function is conceptually wrong because the time factor used does not align with the situation.

164. **Level:** Easy | **Skill/Knowledge:** Nonlinear functions | **Testing Point:** Interpreting value of nonlinear function

Key Explanation: Choice C is correct. The model $f(x)$ describes the revenue of the consulting firm after t years. Since $t = 2$ in $f(2)$, \$40,140.8 represents the revenue of the firm at the end of year 2.

Distractor Explanations: Choice A is incorrect as the statement is false. **Choice B** is incorrect as the statement is false. **Choice D** is incorrect as the statement is false.

165. **Level:** Medium | **Skill/Knowledge:** Nonlinear functions | **Testing Point:** Evaluating meaning of term in nonlinear function

Key Explanation: Choice A is correct. Using the given function, the mass of chemical w after 5 days will be $p(d) = 300(3)^{\frac{5}{5}} = 300(3) = 900$. This new mass is thrice the initial mass of 300. Therefore, the function indicates that the mass of chemical w triples every 5 days.

Distractor Explanations: Choice B is incorrect. The mass does not triple every d days because the time interval d is divided by 5 in the equation. **Choice C** is incorrect. Since the growth factor is 3, the mass triples per time interval. **Choice D** is incorrect. This initial mass of chemical w is 300.

166. **Level:** Easy | **Skill/Knowledge:** Nonlinear functions | **Testing Point:** Determining the meaning in context of terms in nonlinear function

Key Explanation: Choice D is correct. Since the growth factor is 1.12, the growth rate $r = 1.12 - 1 = 0.12$. Hence, the height of the tree increases by 12% annually and not 112%. Therefore, **Choice D** is false.

Distractor Explanations: Choice A is incorrect. This is true statement about the model given. **Choice B** is incorrect. This is true statement about the model given. **Choice C** is incorrect. This is true statement about the model given.

Chapter 5

Problem-Solving and Data Analysis

This chapter includes questions on the following topics:

- Ratios, rates, proportional relationships, and units
- Percentages
- One-variable data: distributions and measures of center and spread
- Two-variable data: models and scatterplots
- Probability and conditional probability
- Inference from sample statistics and margin of error
- Evaluating statistical claims: observational studies and experiments

167

Doug's car holds 50 gallons of fuel. If he drives 150,000 miles at 35 miles per gallon, how many full tanks of gas will he use through his journey if he starts with a full tank?

A) 10

B) 25

C) 72

D) 86

168

If two pounds of grapes cost \$3.96, how much will g pounds of grapes cost?

A) $3.96g$

B) $\left(\frac{3.96}{2}\right)g$

C) $3.96 + g$

D) $\frac{3.96}{g}$

169

There are 36 students at a statistics camp. If two-thirds of the students are girls and three-fourths of the girls are under 5.5 feet, how many girls are under 5.5 feet tall?

170

At Smart Pets, 60% of the 20 dogs have blue eyes. The fraction of hamsters at the store with brown eyes is equal to the fraction of dogs at the store with brown eyes. What is the ratio of hamsters that have brown eyes if the dogs only have brown and blue eyes?

171

12 carpenters take 3 days to build 36 desks for a local school. How many days will it take for 9 carpenters to complete 63 desks if all carpenters work at the same rate?

A) 4

B) 5

C) 6

D) 7

172

A square field has a side length of 2.5 feet. If a fence, which is 36 inches away from either side of the field is created, what is the perimeter of the fence created in inches? (1 foot = 12 inches)

173

Gina's age is 3 times her daughter's age. Five years from now, the ratio of Gina's age to her daughter's age would be 7:3. How old is the daughter now?

Ratios, rates, proportional relationships, and units

174

The ratio of students to teachers in a school is 20.5:k. If there are 164 students in the school and 8 teachers, what is the value of k?

A) 1

B) 2

C) 4

D) 8

175

The price of a school trip per student is inversely proportional to the number of students going on the trip. If 28 students go on the trip, each student pays $$p$. If two students do not attend the trip, what is the new price per student?

A) $\frac{p}{26}$

B) $\frac{14}{13}p$

C) $\frac{13}{14}p$

D) $26p$

176

Martha is a math tutor who helps students prepare for tests. She charges $20 an hour for tutoring students one-on-one, but she charges $25 an hour for pairs of students studying with her in a small group. If Martha has 3 hours to tutor on the night before a test, what would be the difference in her profits if she were to tutor three pairs of students for an hour each vs. three individual students for an hour each (in dollars)?

177

If $\frac{4c}{d}=\frac{1}{4}$, what is the value of $\frac{d}{c}$?

A) $\frac{1}{16}$

B) $\frac{1}{2}$

C) 4

D) 16

178

If $\frac{m}{n}=3$, what is the value of $\frac{6n}{m}$?

A) 0

B) 1

C) 2

D) 6

179

A marine biologist found that the ratio of plant life to animal life in a particular lake is 2:5. If the estimated animal life in the lake is 1,680,000, which of the following best approximates the population of plant life in the lake?

A) 480,000

B) 672,000

C) 1,200,000

D) 4,200,000

180

Mt. Carmel School conducted a study that showed for every three girls in the school, one does not play any sport. 40% of the school's 1,500 population are girls. What is the approximate number of girls that do not play any sport in the school?

181

A biologist models a function for the population of fish in a pond using the equation $f(t) = 1{,}320(2)^t$, where t is the number of years. If the time interval of the function is converted into m months as $f(m) = 1{,}320(2)^{km}$, what is the value of k?

182

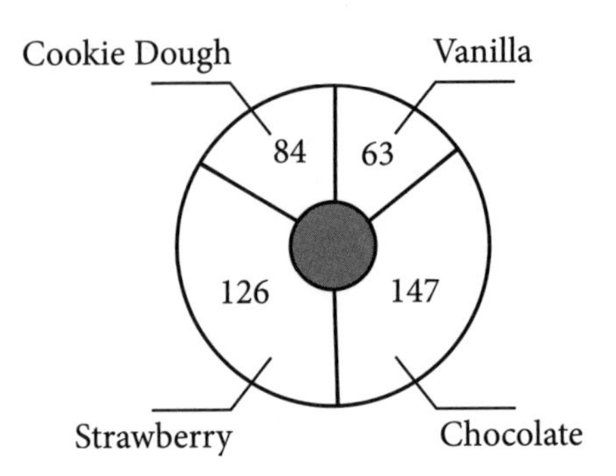

The pie chart above shows the results of a school survey. In the survey, people were asked to choose their favorite ice cream from among a group of four choices. Which of the following was chosen as the favorite ice cream by approximately 35% of the students surveyed?

A) Vanilla

B) Chocolate

C) Strawberry

D) Cookie Dough

167. **Level:** Hard | **Skill/Knowledge:** Ratios, rates, proportional relationships, and units | **Testing Point:** Setting up a proportion and solving problems involving derived units

Key Explanation: Choice D is correct. Create a proportion to express miles per gallon in terms of the total gallons required which yields $\frac{35\,miles}{1\,gallon}=\frac{150,000\,miles}{x\ gallons}$.

Cross multiply to simplify yields $35x = 150,000$. Dividing 35 from both sides of the equation yields
$x = 4,285.71$ gallons.

Divide the total required gallon of fuel by the capacity of the tank to determine how many total tanks are needed. This yields
$\frac{4,285.71}{50}$ or 85.71 tanks.

Rounding to the nearest whole tank yields 86 tanks.

Distractor Explanations: Choice A is incorrect and is most likely the result of rounding error or error in creating and solving proportions. **Choice B** is incorrect and is most likely the result of rounding error or error in creating and solving proportions. **Choice C** is incorrect and is most likely the result of rounding error or error in creating and solving proportions.

168. **Level:** Easy | **Skill/Knowledge:** Ratios, rates, proportional relationships, and units | **Testing Point:** Solving problems involving derived units

Key Explanation: Choice B is correct. If 2 pounds of grapes cost $3.96, then one pound of grapes costs $\frac{3.96}{2}$. Multiplying by g results in the cost of g pounds of grapes, or **Choice B**.

Distractor Explanations: Choice A is incorrect and is the result of an arithmetic error or lack of understanding of ratios and unit costs. **Choice C** is incorrect and is the result of an arithmetic error or lack of understanding of ratios and unit costs. **Choice D** is incorrect and is the result of an arithmetic error or lack of understanding of ratios and unit costs.

169. **Level:** Easy | **Skill/Knowledge:** Ratios, rates, proportional relationships, and units | **Testing Point:** Using proportions to solve word problems

Key Explanation: 18 is correct. Since there are 36 kids in the camp and two-thirds are girls, there are a total of $\frac{2}{3}\times 36=24$ girls at the statistics camp. Of the 24 girls, three-fourths are under 5.5 feet. This means that $\frac{3}{4}\times 24=18$ girls are under 5.5 feet.

170. **Level:** Easy | **Skill/Knowledge:** Ratios, rates, proportional relationships, and units | **Testing Point:** Solving problems involving proportions

Key Explanation: The correct answer is 0.40. To find out the ratio of hamsters that have brown eyes, first determine the fraction of dogs with brown eyes. Since 60% of the 20 dogs have blue eyes and dogs only have brown or blue eyes, then 40% have brown eyes. Therefore, the ratio of hamsters is 40:100 which is 0.40 or $\frac{2}{5}$.

171. **Level:** Hard | **Skill/Knowledge:** Ratios, rates, proportional relationships, and units | **Testing Point:** Working with inverse proportions

Key Explanation: Choice D is correct.
The number of carpenters varies inversely

proportional to the number of days to complete the work. Therefore, $12 \times 3 = 9 \times k$, $k = 4$. This means that it takes 4 days for 9 carpenters to complete 36 desks. $4 \times \frac{63}{36} = 7$ days.

Distractor Explanations: Choice A is incorrect and may be due to conceptual or calculation error. **Choice B** is incorrect and may be due to conceptual or calculation error. **Choice C** is incorrect and may be due to conceptual or calculation error.

172. **Level:** Easy | **Skill/Knowledge:** Ratios, rates, proportional relationships, and units | **Testing Point:** Converting units

Key Explanation: The perimeter of a square field is found by 4 × (length of one side). First, convert all units to inches. The length of one side of the square field is 2.5 which would be $(2.5 \times 12) = 30$ inches. This is the value of one side of the square field. If there is a margin of 36 inches around the field, the length of one side of the fence would be 30 + 36 + 36 inches= 102 inches. The perimeter would therefore be $4(102) = 408$ inches.

173. **Level:** Medium | **Skill/Knowledge:** Ratios, rates, proportional relationships, and units | **Testing Point:** Working with ratios

Key Explanation: Assuming that the daughter's current age is x and Gina's current age is $3x$.

	Now	Five years from now
Gina's age	$3x$	$3x + 5$
Daughter's age	x	$x + 5$

Using the ratio of their ages 5 years from now yields $\frac{3x+5}{x+5} = \frac{7}{3}$. Performing cross multiplication on yields $9x + 15 = 7x + 35$. Subtracting $7x$ and 15 from both sides of the equation yields $2x = 20$. Dividing 2 from both sides of the equation yields $x = 10$. Therefore, Gina's daughter is 10 years old.

174. **Level:** Easy | **Skill/Knowledge:** Ratios, rates, proportional relationships, and units | **Testing Point:** Working with ratios

Key Explanation: Choice A is correct. The ratio of students to teachers is 164:8. Dividing the ratio by 8 yields 20.5:1. Therefore, $k = 1$.

Distractor Explanations: Choice B is incorrect and may be due to conceptual or calculation error. **Choice C** is incorrect and may be due to conceptual or calculation error. **Choice D** is incorrect and may be due to conceptual or calculation error.

175. **Level:** Easy | **Skill/Knowledge:** Ratios, rates, proportional relationships, and units | **Testing Point:** Working with inverse proportions

Key Explanation: Choice B is correct. The total price of the trip can be represented by $\$p \times 28 = \$28p$. Since two students will not attend the trip, 26 students will need to split the total cost equally among themselves. Therefore, the price that each of the 26 students needs to pay would be $\frac{28p}{26}$ which is equivalent to $\frac{14}{13}p$.

Distractor Explanations: Choice A is incorrect. It assumes that p is the total price of the trip. **Choice C** is incorrect and maybe due to conceptual or calculation error. **Choice D** is incorrect and maybe due to conceptual or calculation error.

176. **Level:** Easy | **Skill/Knowledge:** Ratios, rates, proportional relationships, and units | **Testing Point:** Calculating profit from rate data

Key Explanation: If Martha tutors three individual students for one hour each at the rate of $20 an hour, she will make 3 × $20 = $60. Alternatively, if Martha tutors three pairs of students for one hour each at the rate of $25 an hour, she will make 3 × $25 = $75. Therefore, the difference in the two profits is $75 – $60 = $15.

177. **Level:** Medium | **Skill/Knowledge:** Ratios, rates, proportional relationships, and units | **Testing Point:** Rewriting a rational expression

Key Explanation: Choice D is correct. Using cross multiplication, the equation can be rewritten as: $4c \times 4 = d \times 1$.
This simplifies to:

$$16c = d.$$

Dividing both sides of the equation by c results in

$$\frac{d}{c} = 16.$$

Distractor Explanations: Choice A is incorrect and reflects error in simplification. **Choice B** is incorrect and reflects error in simplification. **Choice C** is incorrect and reflects error in simplification.

178. **Level:** Medium | **Skill/Knowledge:** Ratios, rates, proportional relationships, and units | **Testing Point:** Working with equivalent proportions

Key Explanation: Choice C is correct.

If $\frac{m}{n} = 3$, it follows $\frac{n}{m} = \frac{1}{3}$. Since $\frac{n}{m} = \frac{1}{3}$, multiplying 6 to both sides of the equation yields $\frac{6n}{m} = \frac{6}{3}$ or $\frac{6n}{m} = 2$. Therefore, the answer is 2.

Distractor Explanations: Choice A is incorrect and may result from solving the value of $\frac{3n-m}{3m}$. **Choice B** is incorrect and may result from solving the value of $\frac{3n}{m}$. **Choice D** is incorrect and may result from solving the value of $\frac{18n}{m}$.

179. **Level:** Easy | **Skill/Knowledge:** Ratios, rates, proportional relationships, and units | **Testing Point:** Using ratios to find population values

Key Explanation: Choice B is correct. Use the given ratio of 2:5 to create the equation that yields $\frac{Plant\ life}{Animal\ life} = \frac{2}{5}$. Substituting 1,680,000 to the equation yields $\frac{Plant\ life}{1,680,000} = \frac{2}{5}$. Multiplying both sides of the equation by 1,680,000 yields

$$Plant\ life = \frac{2 \times 1,680,000}{5} = 672,000.$$

Distractor Explanations: Choice A is incorrect and is the value of $\frac{2}{7}$ of the animal life in the lake. **Choice C** is incorrect and is the value of $\frac{5}{7}$ of the animal life in the lake. **Choice D** is incorrect and may result due to calculation or conceptual error.

180. **Level:** Easy | **Skill/Knowledge:** Ratios, rates, proportional relationships, and units | **Testing Point:** Using ratios to find a data value

Key Explanation: The correct answer is 200. Since 40% of the school's 1,500 population are girls, there are a total of (40%)(1,500) = 600 girls. Since 1 out of 3 girls does not play any sports, the correct answer is $\frac{1}{3} \times 600 = 200$ girls.

181. **Level:** Medium | **Skill/Knowledge:** Ratios, rates, proportional relationships, and units | **Testing Point:** Modifying an exponential function for a units change

Key Explanation: The correct answer is $\frac{1}{12}$. Since 1 year = 12 months, $t=\frac{m}{12}$. Hence, the function becomes $f(m)=1,320(2)^{\frac{m}{12}}$. Therefore, $k=\frac{1}{12}$.

182. **Level:** Easy | **Skill/Knowledge:** Ratios, rates, proportional relationships, and units | **Testing Point:** Using ratios and percentages to interpret pie chart data results

Key Explanation: Choice B is correct. To find the one that is approximately 35%, determine the ratio of the number of students that chose each flavor using $\frac{\textit{number of students that chose the flavor}}{\textit{total number surveyed}}$. Adding up all the numbers to find the total number of students who participated in the survey yields $63 + 84 + 126 + 147 = 420$. Solving the percentage of students who chose chocolate yields $\frac{147}{420}\times 100\% = 35\%$. Therefore, chocolate is the correct answer.

Distractor Explanations: Choice A is incorrect and is the result of error in creating proportions from a pie chart. **Choice C** is incorrect and is the result of error in creating proportions from a pie chart. **Choice D** is incorrect and is the result of error in creating proportions from a pie chart.

183

There was a sale on sofas at a furniture store for 60% off the original price of any sofa, with an additional 40% off your entire purchase if you buy a matching armchair for a fixed price of $450. Which expression represents the total amount you need to pay if you buy a sofa that is originally $$s$ and a matching armchair?

A) $0.4(0.6s + 450)$

B) $0.6(0.4s + 450)$

C) $0.4s + 0.6(450)$

D) $0.6s + 0.4(450)$

184

A school is renovated so that it is 15% larger than before. If the school currently has a maximum capacity of 744 students after the renovation, what is the maximum capacity (in number of students) it can hold before the renovation (round it to the nearest number)?

185

Number of Questions Correct on Math Exam

Number Correct	Frequency
31–40	3
41–50	6
51–60	7
61–70	9
71–80	3
81–90	4
91–100	2

What percentage of students scored between 0 and 70?

A) 25%

B) 26.5%

C) 73.5%

D) 75%

186

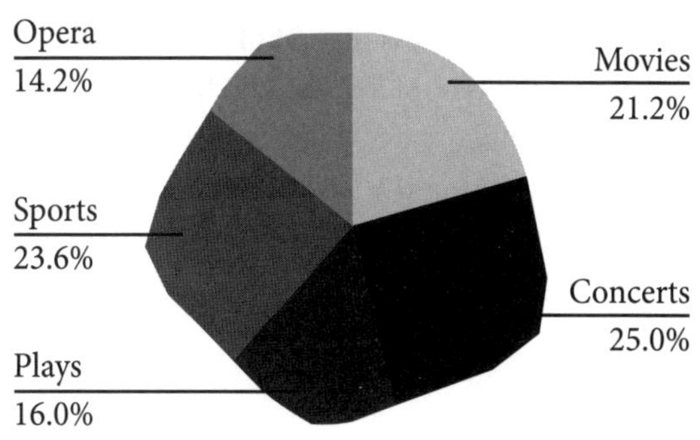

Using the pie chart above, approximately what percent of people did not like sports?

A) 14.2%

B) 23.6%

C) 75%

D) 76.4%

187

A school has a total population of 620 students. The school also sells school supplies to students before the beginning of the year. If 70% of the students purchase school supplies for $25 each, how much does the school make in total?

A) $9,486

B) $10,850

C) $12,543

D) $15,575

188

The price of a dress is $24. If the price is increased by 120%, what is the new price of the dress?

189

The price of a dress in a retail store is $200. During the End of Year Sale at the store, the dress can be bought at a 30% discount. Amanda buys the dress during the End of Year sale and uses a 12% off Coupon. How much does Amanda buy the dress for?

190

The price of fuel in Country X fluctuates. The price decreases by 10% between 2014 and 2016 and then increases by 20% between 2017 and 2019. If the price of fuel in 2019 is $17.28, what was the price of fuel in 2013?

A) $15.55

B) $16.00

C) $18.66

D) $19.00

191

The price of a chocolate bar increases from $0.60 to $2.40. What is the percentage increase in the price of the chocolate bar?

A) 25%

B) 33.33%

C) 300%

D) 400%

192

5 gallons of a 60% salt solution is mixed with x gallons of a 50% salt solution to get a 56.25% salt solution. What is the value of x?

Percentages

193

The price of a winter coat is $180. There's a 15% sales tax imposed on the retail price. Nia bought the coat using a 20% off coupon. How much did she pay for the winter coat?

A) $144.00

B) $165.60

C) $171.00

D) $207.00

194

20% of a positive integer is equal to 60% of 200. What is the value of the positive integer?

A) 24

B) 96

C) 120

D) 600

195

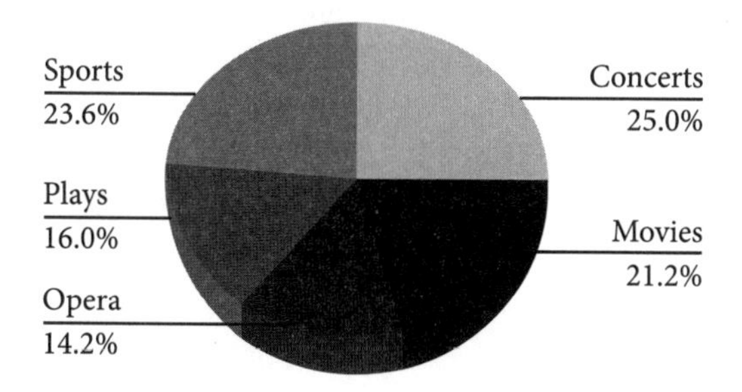

In 2017, approximately 5,500 people were surveyed about preferred weekend activities. How many people preferred opera or sports?

196

A study shows that 1.5% of the houses in the city are commercial houses and the rest are residential houses. If the number of residential houses in the city is 12,000. What is the number of commercial houses in the city? (round off to the nearest integer)

A) 60

B) 150

C) 180

D) 183

183. **Level:** Easy | **Skill/Knowledge:** Percentages
Testing Point: Using percentages to solve problems

Key Explanation: Choice B is correct. In order to correctly represent the situation, match the correct variable with the correct percentage. The sofa s is 60% off, so the total price is in fact 40% of the original price or $0.4s$. The armchair has a fixed price of \$450. The sum of $0.4s$ and \$450 is then discounted again by 40%. So, the total amount to pay is 60% of the sum of $0.4s$ and \$450. Therefore, the correct expression is $0.6(0.4s + 450)$.

Distractor Explanations: Choice A is incorrect because it represents errors in writing discounts as percentages. **Choice C** is incorrect because of error in matching variables and prices with the correct discount. **Choice D** is incorrect because of error in matching variables and prices with correct discount.

184. **Level:** Easy | **Skill/Knowledge:** Percentages
Testing Point: Using percentages to solve problems

Key Explanation: If the size of the school after its renovation is 15% larger than before, then it is now 115% larger. To get the original maximum capacity of the school, divide the current maximum capacity by 115%. This yields $\frac{744}{1.15}$ or 646.95. Rounding to the nearest number yields 647 students.

185. **Level:** Easy | **Skill/Knowledge:** Percentages
Testing Point: Using percentages to solve problems

Key Explanation: Choice C is correct. Adding up the numbers in the frequency column of the table, there are 34 total values. There are 25 out of the 34 scores between 0 and 70. Set up a proportion and solve: $\frac{25}{34} = 73.52941176\%$, which rounds to 73.5%.

Distractor Explanations: Choice A is incorrect and is the result of errors in interpreting frequency chart or calculating percentages. **Choice B** is incorrect and is the result of errors in interpreting frequency chart or calculating percentages. **Choice D** is incorrect and is the result of errors in interpreting frequency chart or calculating percentages.

186. **Level:** Easy | **Skill/Knowledge:** Percentages
Testing Point: Using percentages to solve problems

Key Explanation: Choice D is correct. Since 23.6% of the 100% of people surveyed liked sports, the percentage of people that do not like sports is the difference between the total and those that did like sports. This yields $100\% - 23.6\% = 76.4\%$.

Distractor Explanations: Choice A is incorrect and may result from finding the percentage of people that like operas. **Choice B** is incorrect and may result from finding the percentage of people that like sports. **Choice C** is incorrect and may result from finding the percentage of people that do not like concerts.

187. **Level:** Easy | **Skill/Knowledge:** Percentages
Testing Point: Using percentages to solve problems

Key Explanation: Choice B is correct. First, determine what 70% of the student population is which yields $(0.7)620 = 434$. Then, multiply this value by the price per pack of school supplies which yields $434 \times \$25 = \$10,850$.

Distractor Explanations: Choice A is incorrect and is likely the result of arithmetic error in calculating the percentage of the student population. **Choice C** is incorrect and is likely the result of arithmetic error in calculating the percentage of the student population. **Choice D** is incorrect and is likely the result of arithmetic error in calculating the percentage of the student population.

188. **Level:** Easy | **Skill/Knowledge:** Percentages
Testing Point: Calculating the percentage increase

Key Explanation: $52.80 is the correct answer. The price increases by 120%. Hence, the new price will be 220% of the original price. Multiplying 2.2 to the original price yields 2.2 × 24 or 52.80. The price of the new dress will be $52.80.

189. **Level:** Medium | **Skill/Knowledge:** Percentages
Testing Point: Working with percentage decrease and discount

Key Explanation: During the End of Year Sale, the price of the dress is 30% off which means that it is 70% off its original price. 70% of $200= $140.

Amanda has a further 12% off coupon. This would make the price of the dress to be 88% of $140 = $123.20.

190. **Level:** Hard | **Skill/Knowledge:** Percentages
Testing Point: Using percentage increase and decrease

Key Explanation: Choice B is correct. The price of fuel in 2013 would be the price of fuel before the decrease in 2014. The price in 2016 can be found like this: $\frac{17.28}{1.2}$ = $14.40. The price in 2013 would be $\frac{\$14.40}{0.9}$ = $16 which is the price before the decrease or the price in 2013.

Distractor Explanations: Choice A is incorrect. This would be the price assuming the overall change in price is a 10% increase and $17.28 is the price in 2019. **Choice C** is incorrect. This would be the price in 2019 assuming that $17.28 is the price in 2013. **Choice D** is incorrect. This would be the price assuming the overall change in price is a 10% increase and assuming that $17.28 is the price in 2013.

191. **Level:** Easy | **Skill/Knowledge:** Percentages
Testing Point: Working with percentage increase

Key Explanation: Choice C is correct. The percentage increase is found by $\frac{Increase\ in\ price}{Original\ price} \times 100\%$. In this case, the price of a chocolate bar moves from $0.60 to $2.40. This would yield $\frac{2.40-0.60}{0.60} \times 100\% = \frac{1.80}{0.60} \times 100\% = 300\%$.

Distractor Explanations: Choice A is incorrect. This option may be due to calculation or conceptual error. **Choice B** is incorrect. This option may be due to calculation or conceptual error. **Choice D** is incorrect. This option may be due to calculation or conceptual error.

192. **Level:** Easy | **Skill/Knowledge:** Percentages
Testing Point: Calculating percentages in mixtures

Key Explanation: The correct answer is 3. To solve this, first, calculate the total amount of the resulting mixture by adding the amounts of the two solutions which yield 5 + x. Then, multiply the mass with the corresponding percent solution and add them together which yields 5(0.6) +

$x(0.5)$. Equating this to the resulting mixture yields $5(0.6) + x(0.5) = (5 + x)(0.5625)$. Using distributive property yields $3 + 0.5x = 2.8125 + 0.5625x$. Subtracting $0.5x$ and 2.8125 from both sides of the equation yields $0.1875 = 0.0625x$. Dividing both sides of the equation by 0.0625 yields $3 = x$ or $x = 3$. Therefore, 3 gallons of the 50% salt solution is used in the mixture.

193. **Level:** Medium | **Skill/Knowledge:** Percentages
Testing Point: Using discounts and sales tax percentages

Key Explanation: Choice B is correct. A sales tax of 15% means that the new price is 115% of the retail price. This makes the new price to be $1.15 \times 180 = \$207$. This is the price after the sales tax. Nia used a 20% off coupon which would mean that she pays only 80% of the price which would yield $0.8 \times 207 = \$165.6$.

Distractor Explanations: Choice A is incorrect. This is the price without imposing a sales tax and only applying the discount coupon. **Choice C** is incorrect. This option may be due to misconception or calculation errors. **Choice D** is incorrect. This is the price after the sales tax is imposed and not applying the discount coupon.

194. **Level:** Medium | **Skill/Knowledge:** Percentages
Testing Point: Solving equations involving percentages

Key Explanation: Choice D is correct. Assuming k is the positive integer, the equation will be 20% of k = 60% of 200.

Simplifying the equation yields $0.2k = 120$. Dividing both sides of the equation by 0.2 yields $k = 600$.

Distractor Explanations: Choice A is incorrect. This is the value of 20% of 60% of 200. **Choice B** is incorrect. This is 80% of 60% of 200. **Choice C** is incorrect. This is the value of 60% of 200.

195. **Level:** Easy | **Skill/Knowledge:** Percentages
Testing Point: Using percentages to solve problems

Key Explanation: 2,079 is the correct answer. Find the number of people that prefer opera or sports by multiplying the total percent by the total number of people. This yields (14.2% + 23.6%)(5,500) = 37.8% (5,500) = 0.378 (5,500) = 2,079.

196. **Level:** Medium | **Skill/Knowledge:** Percentages
Testing Point: Using percentages to calculate a data value

Key Explanation: Choice D is correct, The percentage of residential houses in the city is (100 – 1.5)% = 98.5%. 98.5% was represented by 12,000. Therefore, 1.5% of the houses are $\frac{12,000 \times 1.5\%}{98.5\%} = 182.7$, which when rounded off to the nearest integer is 183.

Distractor Explanations: Choice A is incorrect and may be due to conceptual or calculation error. **Choice B** is incorrect and may be due to conceptual or calculation error. **Choice C** is incorrect. This is the value of 1.5% of 12,000.

197

Student Population at Greendale School	
Grade	Number of Students
2nd	74
3rd	73
4th	81
5th	90
6th	75
7th	82

Based on the data in the table above, what grade would the median number of students be in at Greendale School be?

198

The table below represents the number of species of butterfly found in parks over two different seasons.

	Spring	Summer
Park A	32	22
Park B	16	10
Park C	17	20
Park D	40	32
Park E	23	17

What is the difference in the average number of butterflies in the parks per season?

A) 5.4

B) 15.1

C) 20.2

D) 25.1

199

Number of Questions Correct on Math Exam

Number Correct	Frequency
31–40	3
41–50	6
51–60	7
61–70	9
71–80	3
81–90	4
91–100	2

Which measures of central tendency in this set of data have the greater values?

A) Median and Mean

B) Median and Mode

C) Mode and Mean

D) It cannot be determined.

200

A flock of geese flies south for the winter. If their journey is 2,365 miles and they take 3.75 days to complete the journey, what was their average speed in miles per hour? Round your answer to the nearest tenth.

A) 26.3 *mph*

B) 260.2 *mph*

C) 315.3 *mph*

D) 630.7 *mph*

201

$-4, 12, y, 108, -324$

A sequence of 5 integers (which follow a pattern) is shown above, where y represents the median value. What is the average of the first 4 integers?

202

What is the mean for the data set below?

13, 12, 18, 10, 11, 9, 15, 18, 11

A) 9

B) 12

C) 13

D) 18

203

The average score of 30 students on a test is 78%. If the score of the student who scored lowest is removed, what would happen to the average score of the remaining students?

A) The average score of the remaining students remains the same.

B) The average score of the remaining students decreases.

C) The average score of the remaining students increases.

D) There isn't enough information to make a conclusion on the change in the average score.

204

A research assistant collected 11 rocks and weighed them and rounded off their masses to the nearest whole number. The masses of the 11 rocks in kilograms are shown below.

23, 34, 54, 22, 25, 43, 36, 29, 14, 51, 21

If a rock that weighs 11 *kgs* is added to the study, which of the following statements is true about the study?

A) The mean and the range will decrease after the mass is added.

B) The mean and the range will increase after the mass is added.

C) The mean and the range will not change after the mass is added.

D) The mean will decrease but the range will increase after the mass is added.

205

What is the difference between the mean and the median of the data set below?

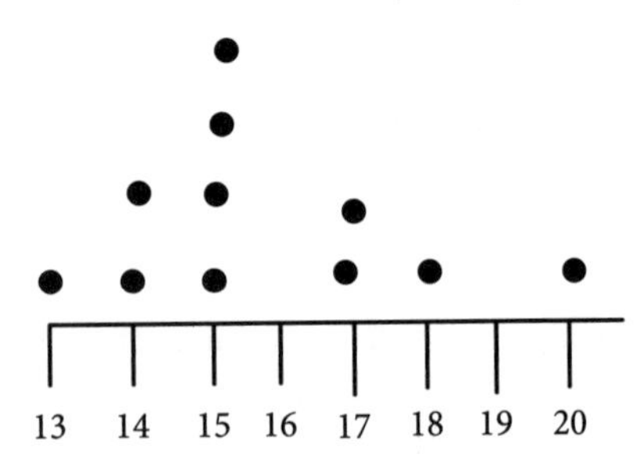

206

Data Set A

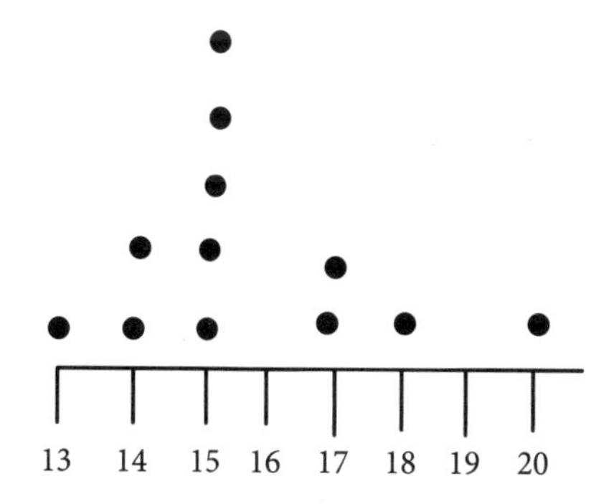

Data Set B

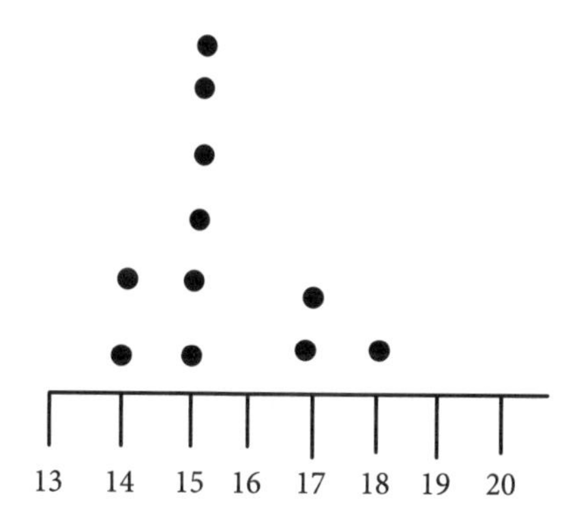

Which of the following statements is true about the two data sets above?

A) The standard deviation of Data Set A is greater than that of Data Set B.

B) The standard deviation of Data Set B is greater than that of Data Set A.

C) The standard deviation of Data Set A is equal to the standard deviation of Data Set B.

D) There isn't enough information to conclude about the standard deviation of the two data sets.

207

What is the mean of the ungrouped data below: 32, 43, 54, 56, 36, 59, 56

208

A math teacher increased his students' scores by 12. Which of the following would increase after this change?

I. Mean
II. Median
III. Range
IV. Standard deviation

A) I and III

B) I and II

C) None of the above

D) All of the above

209

	Class Buck	Class Mow
100–80	1	4
80–60	8	9
60–40	6	2
40–20	1	1

The data above represents how Class Buck and Class Mow performed in their midterm results. Which of the following cannot be true about the two classes represented in the table above?

A) The median of Class Buck and Class Mow is 73.5%.

B) The mean of Class Mow is greater than the mean of Class Buck.

C) The standard deviation of Class Mow is higher than the standard deviation of Class Buck.

D) The median of Class Buck is 57.6%.

197. **Level:** Medium | **Skill/Knowledge:** One-variable data: distributions and measures of center and spread | **Testing Point:** Determining the median value of a set of data

Key Explanation: The correct answer is the 5th grade. To find the median value of the data set, start by finding the total number of students in the school, which yields 74 + 73 + 81 + 90 + 75 + 82 = 475. Since the total number of students is odd, then the expression to get the median is $\frac{n+1}{2}$ where n is the total population. Substituting the data yields $Median = \frac{475+1}{2} = 238$. Therefore, the 238th student from the lowest to the highest grade is the median. Add the number of students per level to accumulate 238. The composition of students will be 74 from 2nd grade, 73 from 3rd grade, 81 from 4th grade, and 10 from 5th grade. Therefore, the 238th student is in the 5th grade.

198. **Level:** Medium | **Skill/Knowledge:** One-variable data: distributions and measures of center and spread | **Testing Point:** Calculating and comparing the means of a set of data

Key Explanation: Choice A is correct. To find the difference between the averages, calculate first the average for each season. This yields $Average_{Spring} = \frac{32+16+17+40+23}{5} = \frac{128}{5} = 25.6$ and

$Average_{Summer} = \frac{22+10+20+32+17}{5} = \frac{101}{5} = 20.2.$

Then, get the difference which yields 25.6 – 20.2 = 5.4.

Distractor Explanations: Choice B is incorrect and may result from a conceptual or calculation error. **Choice C** is incorrect and may result from a conceptual or calculation error. **Choice D** is incorrect and may result from a conceptual or calculation error.

199. **Level:** Hard | **Skill/Knowledge:** One-variable data: distributions and measures of center and spread | **Testing Point:** Comparing and interpreting measures of central tendency

Key Explanation: Choice D is correct. Although from the frequency table the median of the data set can be found and one can estimate the mean, the mode cannot be determined based on a frequency chart.

Distractor Explanations: Choice A is incorrect and is the result of error in interpreting frequency charts. **Choice B** is incorrect and is the result of error in interpreting frequency charts. **Choice C** is incorrect and is the result of error in interpreting frequency charts.

200. **Level:** Easy | **Skill/Knowledge:** One-variable data: distributions and measures of center and spread | **Testing Point:** Finding the average speed from data

Key Explanation: Choice A is correct. Begin by calculating the total time traveled by the geese in hours:

Converting days to hours yields

$3.75\ days \times \frac{24\ hours}{1\ day} = 90\ hours.$

Divide the total distance in miles by the total time in hours to calculate the average speed which yields $Average\ speed = \frac{2,365\ miles}{90\ hours} = 26.2777\ mph.$ Rounding off to the nearest tenth, yields 26.3 mph.

Distractor Explanations: Choice B is incorrect and may result from a conceptual or calculation error. **Choice C** is incorrect and may result from a conceptual or calculation error. **Choice D** is incorrect and may result from a conceptual or calculation error.

One-variable data: distributions and measures of center and spread (Answers)

201. **Level:** Medium | **Skill/Knowledge:** One-variable data: distributions and measures of center and spread | **Testing Point:** Comparing and interpreting mean (average) median range and standard deviation

Key Explanation: The correct answer is 20. Looking at the sequence, each integer is multiplied by –3 to get the following integer. To find y, multiply 12 by –3 which yields $12 \times (-3) = -36$. Therefore, the median is –36. Use this value to find the average of the first four integers which yields $\frac{(-4)+12+(-36)+108}{4} = \frac{80}{4} = 20$.

202. **Level:** Easy | **Skill/Knowledge:** One-variable data: distributions and measures of center and spread | **Testing Point:** Finding the mean of ungrouped data

Key Explanation: Choice C is correct. The mean of ungrouped data can be found by getting the average. This means that the sum of the data will be divided by the number of data. The sum of the data is 13 + 12 + 11 + 10 + 11 + 9 + 15 + 18 + 18 =117. Since there are 9 numbers, the mean would be $\frac{117}{9} = 13$.

Distractor Explanations: Choice A is incorrect. This is the least data value in the set. **Choice B** is incorrect. This is the median of the data set. **Choice D** is incorrect. This is the greatest data value in the set.

203. **Level:** Easy | **Skill/Knowledge:** One-variable data: distributions and measures of center and spread | **Testing Point:** Changing average score measure of central tendency

Key Explanation: Choice C is correct. Removing the lowest score or the outlier on the lower end of the test would increase the average score.

Distractor Explanations: Choice A is incorrect. The average score would not remain the same because removing a data set would lead to a change in the sum of the scores in the test which would in turn lead to a change in the average score. **Choice B** is incorrect. The average score would decrease if a lower data set is added or if the outlier on the higher side is removed. **Choice D** is incorrect; there is enough information to answer the question.

204. **Level:** Easy | **Skill/Knowledge:** One-variable data: distributions and measures of center and spread | **Testing Point:** Finding the mean and the range of data

Key Explanation: Choice D is correct. Adding a data value that would be an outlier on the lower side of the data set would increase the range. The old range was 54 – 14 = 40 while the new range would be 54 – 11 = 43.

The old mean would be 32 and the new mean would be 30.25. The mean has decreased.

Distractor Explanations: Choice A is incorrect as the statement is false. **Choice B** is incorrect as the statement is false. **Choice C** is incorrect as the statement is false.

205. **Level:** Easy | **Skill/Knowledge:** One-variable data: distributions and measures of center and spread | **Testing Point:** Interpreting measures of central tendency from a dot-plot histogram

Key Explanation: The correct answer is $\frac{8}{11}$.

The mean of the data set is given by $\frac{\textit{sum of the values}}{\textit{no. of the values}}$. The sum of the 12 values is $(13 \times 1) + (14 \times 2) + (15 \times 4) + (17 \times 2) + (18 \times 1) + (20 \times 1) = 173$. The mean would therefore be $\frac{173}{11} = 15\frac{8}{11}$ or 15.727. The median of 11 data is given by the average of the 6th data, which would be 15. The difference between the mean and the median would be $\left(15\frac{8}{11}-15\right)=\frac{8}{11}$.

206. **Level:** Medium | **Skill/Knowledge:** One-variable data: distributions and measures of center and spread | **Testing Point:** Finding the standard deviation of grouped data

Key Explanation: Choice A is correct. A standard deviation is a measure of how dispersed the data is with respect to the mean. The mean in Data Set A is $\frac{13+14+14+15+15+15+15+15+17+17+18+20}{12}$ = 15.67. The mean in Data Set B is $\frac{14+14+15+15+15+15+15+15+17+17+18}{11}$ = 15.45. Looking at the charts, the data in Set A is more dispersed from its mean compared to the data in Set B. Therefore, Set A has a greater standard deviation than Set B.

Distractor Explanations: Choice B is incorrect and the statement is false. **Choice C** is incorrect and the statement is false. **Choice D** is incorrect and the statement is false.

207. **Level:** Easy | **Skill/Knowledge:** One-variable data: distributions and measures of center and spread | **Testing Point:** Finding the mean of a group of data

Key Explanation: The correct answer is 48. The mean of the above data is given by the sum of the data set divided by the number of data. The sum of the data is 32 + 43 + 54 + 56 + 36 + 59 + 56 = 336. Therefore, the mean would be $\frac{336}{7}=48$.

208. **Level:** Hard | **Skill/Knowledge:** One-variable data: distributions and measures of center and spread | **Testing Point:** Calculating measures of center and spread

Key Explanation: Choice B is correct. If each data point is increased by 12, the mean and the median would both increase.

Distractor Explanations: Choice A is incorrect. The range of the data set would remain the same because all the data points are increased by the same amount. This would not change the difference between the maximum and minimum data. **Choice C** is incorrect. The mean and range will increase and thus not all options are false. **Choice D** is incorrect. The range and standard deviation remain the same and therefore not all options can be correct.

209. **Level:** Hard | **Skill/Knowledge:** One-variable data: distributions and measures of center and spread | **Testing Point:** Calculating measures of center and spread

Key Explanation: Choice D is correct. This is the only option that is false. Since there are 16 scores in Class Buck, the median is the average of the 8th and 9th scores. The 8th and 9th scores of Class Buck lie between 60 and 80. Therefore, the median cannot be lower than 60 or greater than 80.

Distractor Explanations: Choice A is incorrect. A median of 73.5 is possible since it is between 60 and 80. **Choice B** is incorrect. The mean of Class Mow is between 60 and 80 and the mean of Class Buck is between 51.25 and 71.25. Hence, it is possible that the mean of Class Mow is greater than the mean of Class Buck. **Choice C** is incorrect. It is possible that the standard deviation of Class Mow is greater than the standard deviation of Class Buck.

210

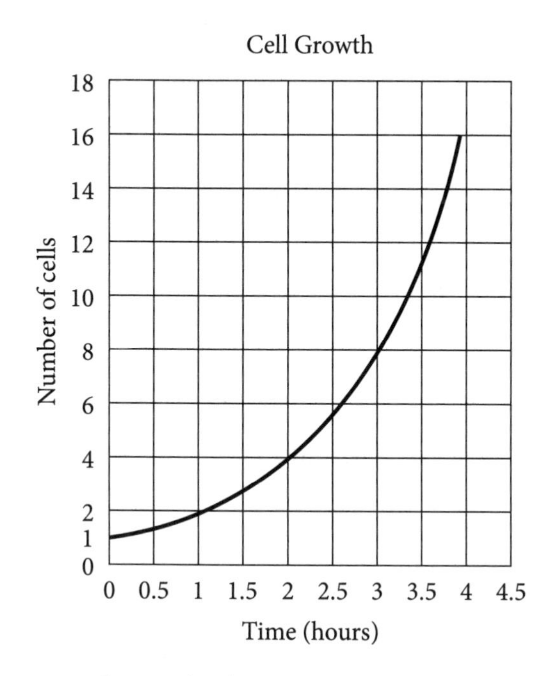

According to the data in the graph of $f(x)$ above, how many hours have passed when $f(x) = 1$?

211

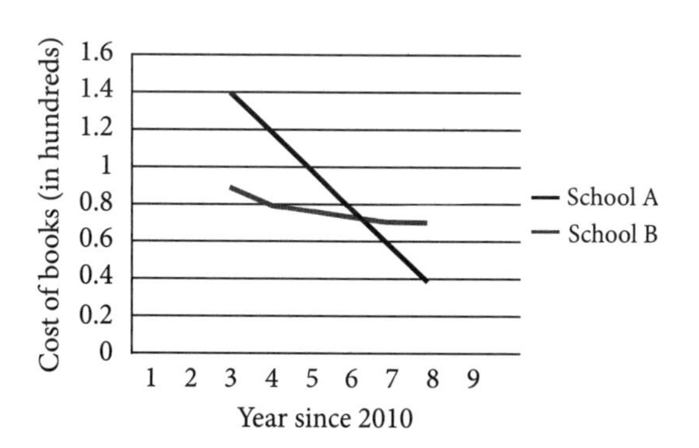

The graph above represents the approximate cost of textbooks in college for two different schools.

Between which years was the cost of textbooks equal at both schools?

A) 2013 – 2014

B) 2014 – 2015

C) 2015 – 2016

D) 2016 – 2017

212

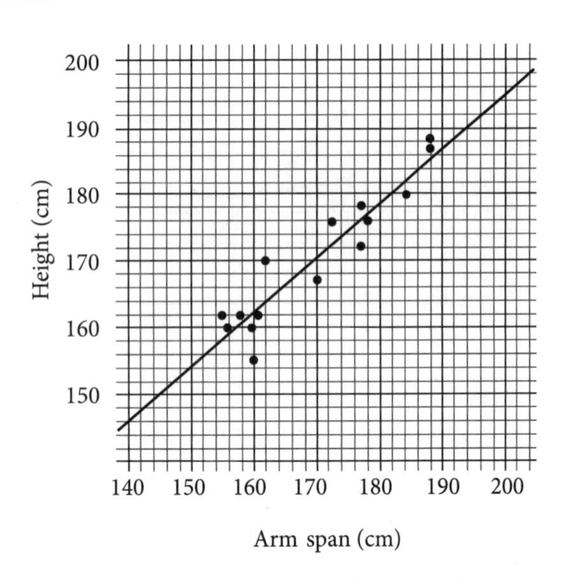

The scatterplot above shows the data collected on the arm span and height of students in a class. A line of best fit with the equation $y = 0.85x + 25.68$ where x is arm span in *cm* and y is height in *cm*, is calculated for the data. Based on the line of best fit, if the arm span of a student is 166 centimeters, what is the predicted height in centimeters?

A) 163.64

B) 166.78

C) 168.91

D) 169.21

213

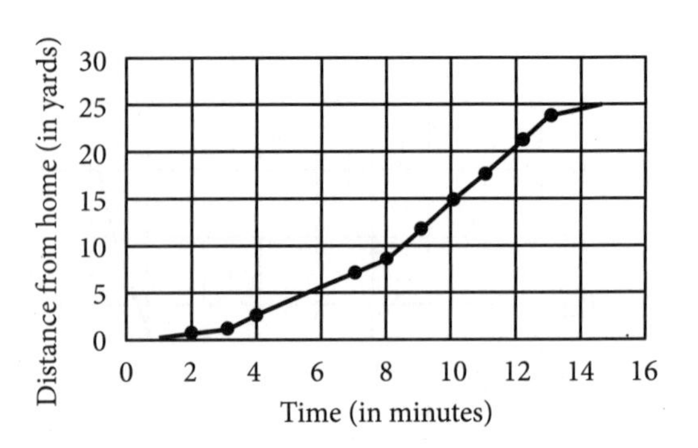

The scatterplot above represents the distance that Janet has walked going to her friend's house from her home. According to the graph, at approximately what time interval does Janet walk the fastest?

A) 0 – 2 minutes

B) 4 – 6 minutes

C) 8 – 10 minutes

D) 12 – 14 minutes

214

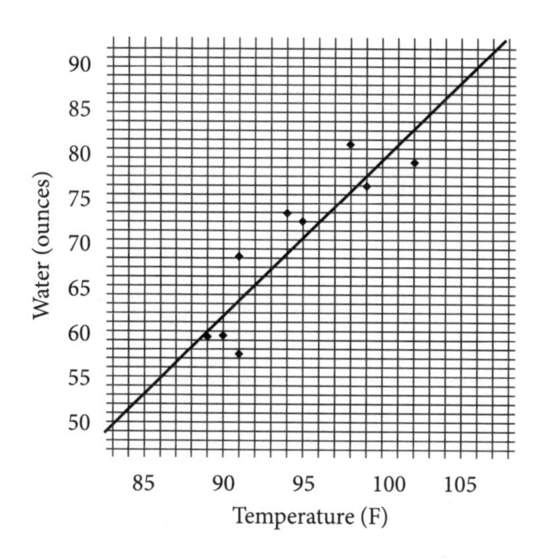

The scatterplot above shows the relationship between the outdoor temperature and the water consumption for a camper on a 10-day camping trip. One day is not included in the scatterplot where the temperature was 97° F. Based on the line of best fit, which of the following is closest to the amount of water consumed by the camper on that day?

A) 68 ounces

B) 70 ounces

C) 75 ounces

D) 82 ounces

215

The temperatures y in a town on x days were recorded °C and plotted on the graph below. Using the line of best fit as the predicted value, by how many °C is the predicted value on day 4 greater than the actual value recorded?

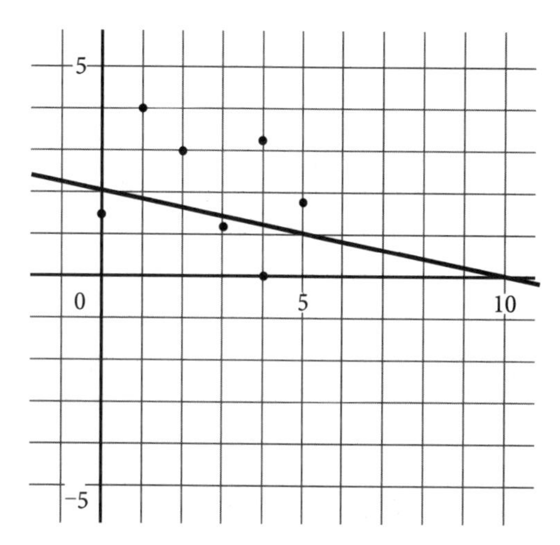

216

The temperature in town A is recorded for ten days and represented in the scatterplot below.

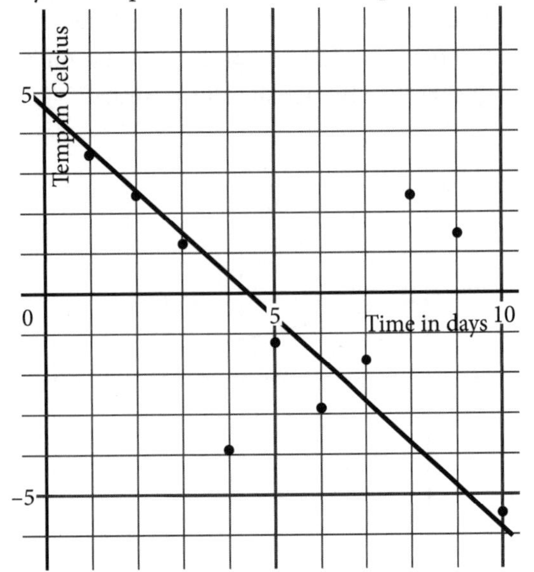

For how many days actual temperatures does the line of best fit predict a higher temperature?

217

Which of the following best describes the equation of the line of best fit?

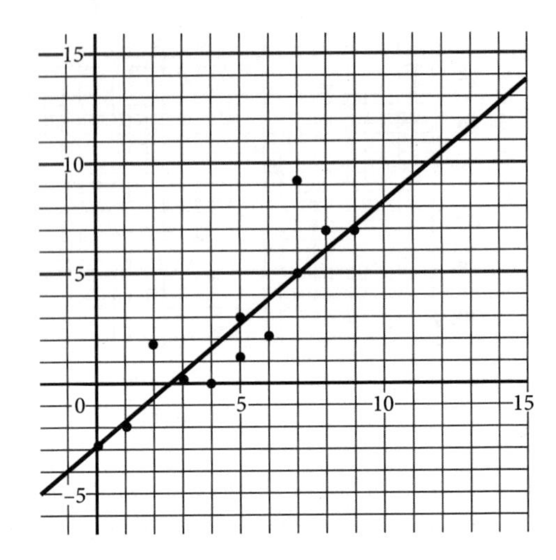

A) $y = -3x + 2.8$

B) $y = 1.1x - 2.8$

C) $y = 2.1x - 2.4$

D) $y = -1.2x - 2.8$

218

Which of the following is the equation that best represents the graph below?

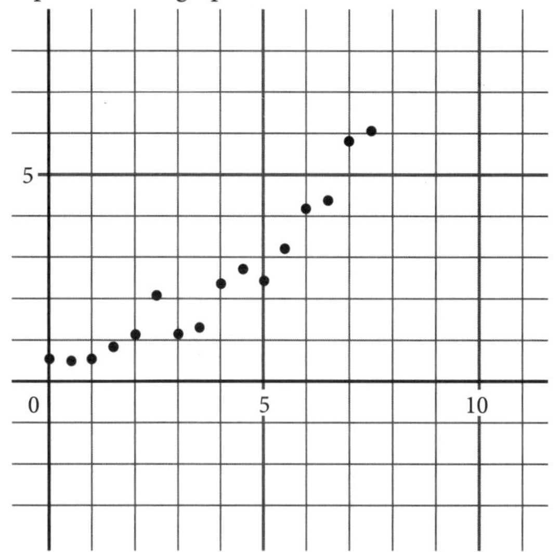

A) $y = 3(1.5)^x$

B) $y = 0.85(1.2)^{2x}$

C) $y = 0.85(0.2)^{2x}$

D) $y = 1.2(0.85)^{2x}$

219

The graph below represents the model $f(x)$. What is the value of $f(3.5)$ that is predicted by the line of best fit?

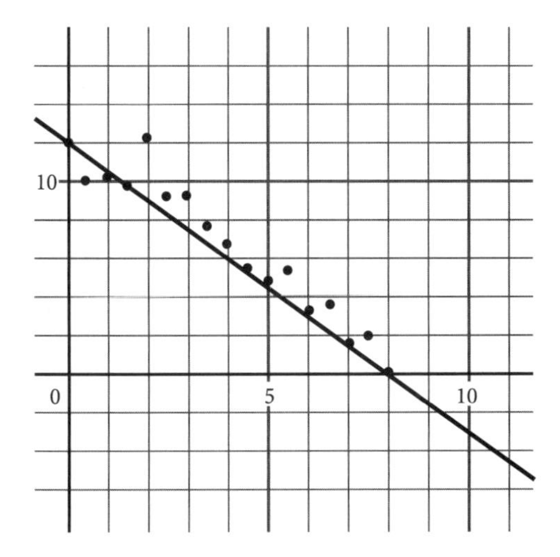

220

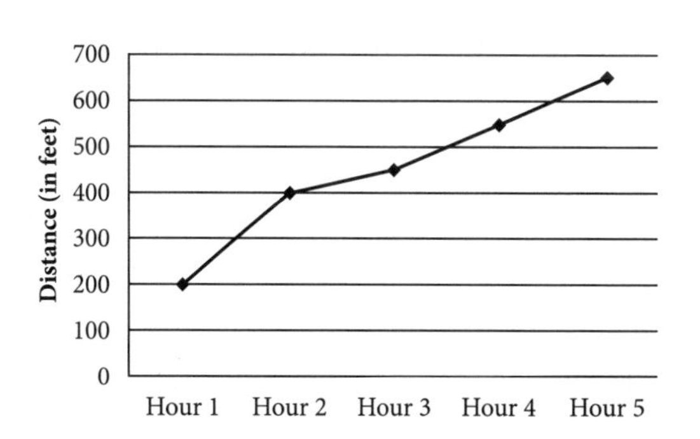

Lisa and Vivian hike a trail one afternoon. The graph above shows the distance they traveled over time. At which interval do Lisa and Vivian hike at the slowest speed?

A) Between hours 1–2

B) Between hours 2–3

C) Between hours 3–4

D) Between hours 4–5

221

Number of Points	10	20	30	40	50	60	70	80	90	100
Jerome	2	1	0	1	3	0	0	1	1	0
Matilda	4	0	0	3	1	0	0	1	0	0

Jerome and Matilda are playing darts. There are 10 different point sections on the board and Jerome and Matilda have already played 9 of their 10 darts in this game. The information is represented in the table above. If Jerome hits the 70-point spot on his final dart, what must Matilda score to beat Jerome?

A) 80

B) 90

C) 100

D) Not possible given her current score

222

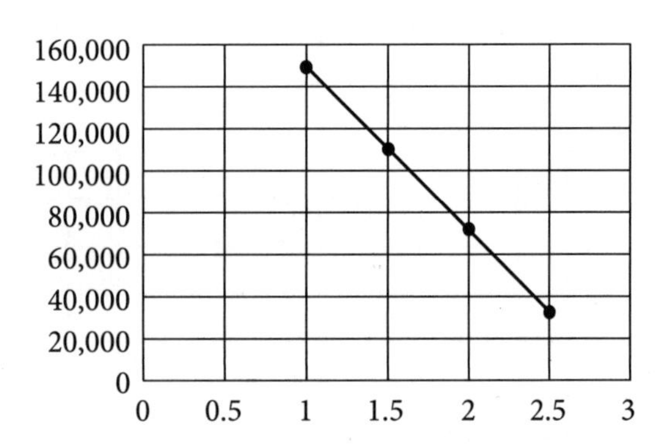

The graph above represents the average salary as it corresponds to the distance from the city center. According to the graph, which value is closest to the average salary at 1.75 miles from the city center?

A) \$60,000

B) \$70,000

C) \$90,000

D) \$120,000

223

Year	Driver's Licenses
2000	141,000
2002	137,000
2005	133,000
2006	127,000
2011	102,000
2014	91,800

Using the table above, a regression analysis between the number of driver's licenses issued in Humble, Texas and the year resulted in the following equation: $y = -3{,}725x + 7{,}596{,}500$.

According to the regression equation:

A) There is a decrease in the number of driver's licenses issued since the year 2000.

B) There is an increase in the number of driver's licenses issued since the year 2000.

C) There is no evidence of an increase or decrease in the number of driver's licenses issued since the year 2000.

D) There is no decrease or increase in the number of driver's licenses issued since the year 2000.

210. **Level:** Easy | **Skill/Knowledge:** Two-variable data: models and scatterplots | **Testing Point:** Solving problems involving two variable graphs

Key Explanation: The graph represents an exponential function. Looking at the graph, $y = 1$ at the start or when $x = 0$. Therefore, the number of cells is initially 1 and zero hours have passed.

211. **Level:** Easy | **Skill/Knowledge:** Two-variable data: models and scatterplots | **Testing Point:** Interpreting data on two-variable graphs

Key Explanation: Choice D is correct. The point where the two graphs intersect is where the cost of the textbooks is the same. The intersection point lies between the 6th and 7th year after 2010. Therefore, the answer is between 2016 and 2017.

Distractor Explanations: Choice A is incorrect and may result from an error in interpreting the data on the graph. **Choice B** is incorrect and may result from an error in interpreting the data on the graph. **Choice C** is incorrect and may result from an error in interpreting the data on the graph.

212. **Level:** Easy | **Skill/Knowledge:** Two-variable data: models and scatterplots | **Testing Point:** Solving problems using line of best fit on a scatterplot

Key Explanation: Choice B is correct. To find the predicted height, substitute 166 for x in the line of best fit equation as follows:

$$y = 0.85x + 25.68$$
$$y = 0.85(166) + 25.68 = 166.78.$$

Distractor Explanations: Choice A is incorrect and reflects error in simplification or misinterpretation of the line of best fit. **Choice C** is incorrect and reflects error in simplification or misinterpretation of the line of best fit. **Choice D** is incorrect and reflects error in simplification or misinterpretation of the line of best fit.

213. **Level:** Medium | **Skill/Knowledge:** Two-variable data: models and scatterplots | **Testing Point:** Reading a scatterplot graph

Key Explanation: Choice C is correct. Speed can be calculated as $\frac{yards}{minute}$, which on this graph is $\frac{y}{x}$ or the slope. The fastest speed is therefore where the slope is the largest or steepest. And the steepest line among the choices is between minutes 8 and 10. The answer is, therefore, **Choice C**.

Distractor Explanations: Choice A is incorrect because the speed at those time intervals is less than the speed at 8–10 minutes. **Choice B** is incorrect because the speed at those time intervals is less than the speed at 8–10 minutes. **Choice D** is incorrect because the speed at those time intervals is less than the speed at 8–10 minutes.

214. **Level:** Easy | **Skill/Knowledge:** Two-variable data: models and scatterplots | **Testing Point:** Solving problems involving line of best fit

Key Explanation: Choice C is correct. Identify $97° F$ on the x-axis. Then identify the intersection on the line of best fit which is at $y = 75$.

Distractor Explanations: Choice A is incorrect and is the result of errors in interpreting lines of best fit. **Choice B** is incorrect and is the result of errors in interpreting lines of best fit. **Choice D** is incorrect and is the result of errors in interpreting lines of best fit.

215. **Level:** Medium | **Skill/Knowledge:** Two-variable data: models and scatterplots | **Testing Point:** Finding the equation of the line of best fit

Key Explanation: The actual temperatures recorded on day 4 were 0 and 3.1. Looking at the graph, the x and y intercepts of the line of best fit are (10, 0) and (0, 2), respectively. Solving the slope of the line yields $\frac{2-0}{0-10} = -\frac{1}{5}$. Hence, the equation of the line of best fit is $y = -\frac{1}{5}x + 2$. The predicted value of 4th day would be $y=-\frac{1}{5}(4)+2$ or $y=-\frac{4}{5}+2$. Simplifying the equation yields $y=\frac{6}{5}$ or y = 1.2. The value that is less than 1.2 is 0. Therefore, the difference is 1.2.

216. **Level:** Easy | **Skill/Knowledge:** Two-variable data: models and scatterplots | **Testing Point:** Interpreting scatterplots and line of best fit

Key Explanation: The correct answer is 6. To solve this, count the number of data below the line of best fit. There are 6 points below the line.

217. **Level:** Easy | **Skill/Knowledge:** Two-variable data: models and scatterplots | **Testing Point:** Determining the equation of line of best fit from graph

Key Explanation: Choice B is correct. Since the graph rises from left to right, then the slope of the line of best fit is positive. Looking at the choices, only **Choices B** and **C** have positive slopes indicated by the positive value of the coefficient of x. Hence, one of them is the correct answer.

To determine which of the two choices is correct, solve the x-intercept of each equation by substituting 0 to y.

Hence, **Choice B** becomes $0 = 1.1x - 2.8$. Adding 2.8 and dividing both sides of the equation by 1.1 yields $x=\frac{2.8}{1.1}=2.55$.

Getting the x-intercept of **Choice C** yields $0 = 2.1x - 2.4$. Adding 2.4 and dividing both sides of the equation by 2.1 yields $x=\frac{2.4}{2.1}=1.14$.

Since the x-intercept of **Choice B** matches the x-intercept in the graph, $y = 1.1x - 2.8$ is the equation of the line of best fit.

Distractor Explanations: Choice A is incorrect because the equation has a negative value of the slope. **Choice C** is incorrect because the x-intercept of the line is not matched with the graph. **Choice D** is incorrect because the equation has a negative value of the slope.

218. **Level:** Hard | **Skill/Knowledge:** Two-variable data: models and scatterplots | **Testing Point:** Determining the exponential equation from graph

Key Explanation: Choice B is correct. The graph is an increasing exponential graph. An exponential growth function has an equation of $y = a(1 + r)^x$ where a is the y-intercept of the graph and $(1 + r)$ is the growth factor. The y-intercept of the graph is less than 1. Hence, the answer can only be **Choice B** or **Choice C**. Since the growth factor of **Choice B** is greater than 1, $y = 0.85(1.2)^{2x}$ is the equation of the graph.

Distractor Explanations: Choice A is incorrect because the y-intercept in the equation is greater than 1. **Choice C** is incorrect because a growth factor of less than 1 implies that the equation is an exponential decay function. **Choice D** is incorrect because the y-intercept in the equation is greater than 1.

219. **Level:** Easy | **Skill/Knowledge:** Two-variable data: models and scatterplots | **Testing Point:** Finding the predicted value from the line of best fit

Key Explanation: To find the equation of the line of the best fit, first find the slope. Using the points (0, 12) and (8, 0) to find the slope yields $\frac{12-0}{0-8}=-\frac{12}{8}=\frac{-3}{2}$. The equation of the line of best fit can be written in the slope-intercept form $y = mx + c$, where c is the y-intercept and m is the slope. Hence, the equation of the line of best fit is $y = -1.5x + 12$. Substituting 3.5 to x and $f(3.5)$ to y in the equation yields $f(3.5) = -1.5(3.5) + 12 = 6.75$.

220. **Level:** Easy | **Skill/Knowledge:** Two-variable data: models and scatterplots | **Testing Point:** Interpreting data values on a graph and the distance between them

Key Explanation: Choice B is correct. To determine at which interval Lisa and Vivian hiked at the slowest speed, we can simply observe each hour's slope because it represents the rate of change of distance with respect to time. A lower (or less steep) slope value indicates a slower speed, as it means less distance was covered over a unit of time.
From the graph, the slowest speed occurs between hours 2–3, where the change in distance is the smallest over one hour.

Distractor Explanations: Choice A is incorrect and is most likely the result of error in understanding and interpreting the data on a graph. **Choice C** is incorrect and is most likely the result of error in understanding and interpreting the data on a graph. **Choice D** is incorrect and is most likely the result of error in understanding and interpreting the data on a graph.

221. **Level:** Easy | **Skill/Knowledge:** Two-variable data: models and scatterplots | **Testing Point:** Evaluating data models

Key Explanation: Choice D is correct. After 9 darts, Jerome has a total of 400 points and Matilda has a total of 290 points. If Jerome throws a 70-point throw, his score increases to 470 points. Because the difference between their two scores is over 100 points, which is the largest single point value, it is therefore impossible for Matilda to beat Jerome with only one more throw.

Distractor Explanations: Choice A is incorrect because every value added to 290 is still lesser than 470 points. **Choice B** is incorrect because every value added to 290 is still lesser than 470 points. **Choice C** is incorrect because every value added to 290 is still lesser than 470 points.

222. **Level:** Easy | **Skill/Knowledge:** Two-variable data: models and scatterplots | **Testing Point:** Interpreting a value from a graph

Key Explanation: Choice C is correct. To determine the average salary (y), when (x) is 1.75, use the line of best fit. The line of best fit includes the ordered pair (1.75, 90,000), so the average income is approximately $90,000.

Distractor Explanations: Choice A is incorrect and is the result of incorrectly interpreting the data on the graph. **Choice B** is incorrect and is the result of incorrectly interpreting the data on the graph. **Choice D** is incorrect and is the result of incorrectly interpreting the data on the graph.

223. **Level:** Easy | **Skill/Knowledge:** Two-variable data: models and scatterplots | **Testing Point:** Calculating trends in table and working with algebraic equation

Key Explanation: Choice A is correct. Looking at the data on the graph as the year increases, there is a trend showing that the number of driver's licenses issued is decreasing. Additionally, the regression equation is written in the form of $y = mx + b$ where m is the slope and b is the y-intercept. In the regression equation $y = -3{,}725x + 7{,}596{,}500$ and thus the slope is $-3{,}725$ which is negative and thus indicates that as the years increase, the number of driver's licenses decrease.

Distractor Explanations: Choice B is incorrect and is the result of interpreting trends in table data incorrectly or the slope of a linear function incorrectly. **Choice C** is incorrect and is the result of interpreting trends in table data incorrectly or the slope of a linear function incorrectly. **Choice D** is incorrect and is the result of interpreting trends in table data incorrectly or the slope of a linear function incorrectly.

224

Mary has a jar of different flavored candies. She has 47 total pieces, including 16 chocolates and 17 peppermint-flavored candies. If she also has a third type of candy, which are caramel flavored, what is the probability that Mary will NOT choose a caramel candy from the jar?

A) $\frac{14}{47}$

B) $\frac{16}{47}$

C) $\frac{17}{47}$

D) $\frac{33}{47}$

225

	Took a gap year	Did not take a gap year	Total
Freshman	37	75	112
Sophomore	21	62	83
Junior	40	60	100
Senior	15	82	97
Total	113	279	392

The table above includes the results of a survey of students at a school who took a gap year.

If a student is randomly chosen from the freshman and sophomore, what is the probability that the student took a gap year (round it to the nearest tenth decimal)?

226

	0–5 hours	6–10 hours	11–15 hours	Total students
Grade 10	21	16	25	62
Grade 11	15	17	35	67
Total students	36	33	60	

According to the data in the table above, if one was to choose a student at random from grade 10, what is the probability that the student studied at least 6 hours a night?

227

	Declared Pre-Med Majors	Non-Majors	Total
Passed Biology 101	204	150	354
Did Not Pass Biology 101	32	55	87
Total	236	205	441

The table above shows the pass/fail results of a class of Biology 101 students, which includes Pre-Med majors and some Non-Major students.

If a single student is chosen from the full class, what is the probability that they are a Non-Major student who passed the exam?

A) $\frac{32}{441}$

B) $\frac{87}{441}$

C) $\frac{150}{441}$

D) $\frac{204}{441}$

228

In a carnival game, there are a total of n bean bags in a sack. The bean bags are either red or yellow. If the probability of choosing a red bean bag at random from the sack is 60%, which of the following represents the number of red bean bags in the sack bags?

A) $\frac{3n}{5}$

B) $6n$

C) $\frac{5n}{3}$

D) $60n$

229

The chance that it would rain in the first two weeks of December is 50%. If it rains on the first 3 days of December, what would be the probability that it would also rain on the 4th day of December?

A) $\frac{3}{14}$

B) $\frac{4}{14}$

C) $\frac{7}{14}$

D) $\frac{12}{14}$

230

A restaurant has 102 items on the menu. The menu contains either dairy-free or gluten-free items and these items either contain nuts or not.

	Gluten-free	Dairy-free	Total
With nuts		32	50
Without nuts			
	48		102

What is the probability that a customer chooses a dairy-free meal without nuts?

A) $\frac{11}{51}$

B) $\frac{16}{51}$

C) $\frac{11}{26}$

D) $\frac{26}{51}$

231

Seven balls numbered 1 to 7 are inside a box. If a ball is randomly picked from the box, what is the probability that it is a factor of 6?

232

A chocolate box has 3 flavors: Mango, White, and Dark chocolate. There are 12 pieces of each flavor inside the box. Marie takes a mango-flavored chocolate. What is the probability that she would pick a mango-flavored chocolate a second time?

A) $\frac{11}{36}$

B) $\frac{11}{35}$

C) $\frac{1}{3}$

D) $\frac{4}{11}$

233

A fair coin is flipped 9 times. It lands on tails 5 times and on heads 4 times. What is the probability of flipping a head on the 10th time?

234

A bag has p blue-only marbles, g red-only marbles, and f marbles that are both red and blue. What is the probability that Amber picks a marble that has the color red?

A) $\frac{g+f}{g+f+p}$

B) $\frac{p+f}{g+f+p}$

C) $\frac{p}{g+f+p}$

D) $\frac{g}{g+f+p}$

235

The probability that Goldie carries an umbrella during a certain week is 0.45. During that week, the probability that it would not rain was 0.2. What is the probability that Goldie carries an umbrella on a rainy day?

A) 0.36

B) 0.45

C) 0.55

D) 0.8

224. **Level:** Hard | **Skill/Knowledge:** Probability and conditional probability | **Testing Point:** Using the concept of probability to solve problems

Key Explanation: Choice D is correct. The probability that Mary will not choose a caramel-flavored candy can also be expressed as the probability that Mary will choose a chocolate or a peppermint, which can be expressed as $\frac{16}{47}+\frac{17}{47}=\frac{33}{47}$.

Distractor Explanations: Choice A is incorrect because it represents the probability that Mary will choose a caramel candy. **Choice B** is incorrect because it represents the probability that Mary will choose a chocolate candy. **Choice C** is incorrect because it represents the probability that Mary will choose a peppermint candy.

225. **Level:** Medium | **Skill/Knowledge:** Probability and conditional probability | **Testing Point:** Using the concept of probability to solve problems

Key Explanation: The total number of students in freshman and sophomore that could be chosen is 112 + 83 = 195. The number of students in freshman and sophomore who took a gap year is 37 + 21 = 58. Therefore, the probability of choosing a student who took a gap year is $\frac{58}{195}$ = 0.297 = 0.3 (rounded to the nearest tenth).

226. **Level:** Medium | **Skill/Knowledge:** Probability and conditional probability | **Testing Point:** Using definition of probability to solve problem

Key Explanation: $\frac{41}{62}$ or approximately 0.661 is correct. If you are choosing a student at random from Grade 10, the total number of students to choose from is 62. If a student has studied at least 6 hours, then he or she has studied anywhere between 6 and 15 hours, so the probability can be expressed as $\frac{25+16}{62}=\frac{41}{62}$.

227. **Level:** Easy | **Skill/Knowledge:** Probability and conditional probability | **Testing Point:** Using the concept of probability to solve problems

Key Explanation: Choice C is correct. The probability is a fraction of the total class, so the denominator of the fraction must be 441. Next, identify which value on the table represents Non-Majors who passed the exam, which is 150. The probability is therefore $\frac{150}{441}$.

Distractor Explanations: Choice A is incorrect and may result from finding the probability to pick a Pre-Major student who did not pass. **Choice B** is incorrect and may result from finding the probability to pick a Pre-Major or Non-Major student who did not pass. **Choice D** is incorrect and may result from finding the probability to pick a Pre-Major student who passed.

228. **Level:** Medium | **Skill/Knowledge:** Probability and conditional probability | **Testing Point:** Using the concept of probability to solve problems

Key Explanation: Choice A is correct. Since $60\%=\frac{60}{100}=\frac{3}{5}$, then $\frac{3}{5}$ of the total bean bags will be red. Therefore, multiply the total number of bean bags n by $\frac{3}{5}$ which results in $\frac{3n}{5}$.

Distractor Explanations: Choice B is incorrect and may result from a conceptual or calculation error. **Choice C** is incorrect and may result from a conceptual or calculation error. **Choice D** is incorrect and may result from a conceptual or calculation error.

229. **Level:** Easy | **Skill/Knowledge:** Probability and conditional probability | **Testing Point:** Finding probability of an event

Key Explanation: Choice C is correct. The probability that it would rain in the first two weeks of December is 50% which means the probability that it would rain on the fourth day is also 50%. This is given by the fraction $\frac{7}{14}$.

Distractor Explanations: Choice A is incorrect and would be the result of a conceptual or calculation error. **Choice B** is incorrect and would be the result of a conceptual or calculation error. **Choice D** is incorrect and would be the result of a conceptual or calculation error.

230. **Level:** Easy | **Skill/Knowledge:** Probability and conditional probability | **Testing Point:** Calculating probabilities

Key Explanation: Choice A is correct. Probability is found by favorable outcomes/ sample space. First complete the table:

	Gluten-free	Dairy-free	Total
With nuts	18	32	50
Without nuts	30	22	52
	48	54	102

The number of dairy-free products without nuts is 22. Therefore, the probability that a customer would choose a dairy-free meal without nuts would be $\frac{22}{102}=\frac{11}{51}$.

Distractor Explanations: Choice B is incorrect. This is the probability of randomly choosing a dairy-free item with nuts. **Choice C** is incorrect. This is the probability of choosing a dairy-free meal from products without nuts. **Choice D** is incorrect. This is the probability of randomly choosing an item without nuts.

231. **Level:** Medium | **Skill/Knowledge:** Probability and conditional probability | **Testing Point:** Calculating probabilities

Key Explanation: The correct answer is $\frac{4}{7}$. The factors of 6 are 1, 2, 3, and 6. Therefore, the probability that the numbered ball is a factor of 6 is $\frac{4}{7}$.

232. **Level:** Hard | **Skill/Knowledge:** Probability and conditional probability | **Testing Point:** Calculating conditional probabilities

Key Explanation: Choice B is correct. After picking the first chocolate which was mango-flavored, there were 11 mango flavors left. Consequently, there were $36 - 1 = 35$ chocolates left in the box. The probability that Marie would pick the mango flavor for the second time would be $\frac{11}{35}$.

Distractor Explanations: Choice A is incorrect and may result from not considering the number of chocolates left in the box after the first pick. **Choice C** is incorrect. This is the probability of picking any flavour for the first time. **Choice D** is incorrect and may result from a conceptual or calculation error.

233. **Level:** Easy | **Skill/Knowledge:** Probability and conditional probability | **Testing Point:** Calculating probability of independent events

Key Explanation: The correct answer is 0.5. The probability of landing heads in a fair coin is always 0.5 regardless of the number of times the coin was flipped.

234. **Level:** Medium | **Skill/Knowledge:** Probability and conditional probability | **Testing Point:** Calculating probabilities

Key Explanation: Choice A is correct. The total number of marbles that have red is $g + f$. The total number of marbles in the bag is $g + f + p$. Therefore, the probability of picking a marble with red color is $\frac{g+f}{g+f+p}$.

Distractor Explanations: Choice B is incorrect. This is the probability of randomly picking a marble that has the color blue. **Choice C** is incorrect. This is the probability of randomly picking a marble that only has the color blue. **Choice D** is incorrect. This is the probability of randomly picking a marble that only has the color red.

235. **Level:** Easy | **Skill/Knowledge:** Probability and conditional probability | **Testing Point:** Calculating complement probabilities

Key Explanation: Choice A is correct. The probability that it would rain can be calculated as $(1 - 0.2) = 0.8$. The probability that Goldie carries an umbrella on a rainy day is $0.45 \times 0.8 = 0.36$.

Distractor Explanations: Choice B is incorrect. This is the probability that Goldie carries an umbrella. **Choice C** is incorrect as this is the probability that Goldie does not carry an umbrella. **Choice D** is incorrect. This is the probability that it would rain that week.

Inference from sample statistics and margin of error

236

An IT company selected 1,000 employees randomly. If $p = 0.9$, or 90% were in favor of 4-day-weeks. The standard error of p (the sample proportion) rounded to the nearest thousandth is given by?

A) 0.009

B) 0.049

C) 0.064

D) 0.081

237

In a random sample of 300 widgets, 4 are defective. Using this data, how many of the 12,000 widgets would one estimate to be defective?

238

A random sample of 10,000 employees was asked whether they had the habit of smoking. Fourteen percent of the 10,000 employees said they had. Which one of the following statements about the number 0.14 is correct?

A) It is a sample proportion.

B) It is a population proportion.

C) It is a margin of error.

D) It is a randomly chosen number.

239

A botanist found that the average height of a Moringa tree is 34.6 feet. The estimated margin of error in the study was 1.2 feet. Which of the following statements best represents the information given?

A) The trees have an average height of 35.68 feet.

B) All Moringa trees have a height between 33.4 feet and 35.8 feet.

C) The average height of a Moringa tree lies between 33.4 feet and 35.8 feet.

D) All Moringa trees in the study have a height between 33.4 feet and 35.8 feet.

240

The average weight of high school students in a class is 147.8 *lbs*. The maximum average weight of the high school students was 160 *lbs* and the minimum average weight is 135.6 *lbs*. What is the margin of error associated with the average weight of high school students?

241

A study was conducted on 20 students at an SAT prep center to find the average SAT score. Which of the following statements would reduce the margin of error?

A) Increasing the sample of students from the SAT test prep center

B) Increasing the sample of students from students at a nearby SAT prep center

C) Decreasing the sample size of students from the SAT test prep center

D) Conducting the study with the top ten students at the SAT test prep center

242

A representative sample of households in City X are surveyed. The survey shows that 17.8% of the households in the sample have cars. If City X has a total of 21,000 households, approximately how many of them have cars?

A) 1,869

B) 2,100

C) 3,738

D) 7,476

243

A study was conducted on how many teachers in an elementary school in City A had a postgraduate diploma. The teachers that participated in the study were chosen randomly. The study showed that 28 out of the 35 teachers have a postgraduate diploma. Which of the following can best be concluded from the given information?

A) 80% of teachers in City A have a postgraduate diploma.

B) 80% of teachers in cities have a postgraduate diploma.

C) 80% of teachers in elementary school have a postgraduate diploma.

D) Nothing can be concluded from the study.

244

100 residents in Town X were surveyed. 56% of the respondents voted yes to building a park in the middle of the town. The town's population is 60,000. Which of the following is the best conclusion drawn from the given information?

A) An estimated 26,400 of the population of Town X would vote yes to building a park in the middle of the town.

B) An estimated 33,600 of the population of Town X would vote yes to building a park in the middle of the town.

C) An estimated 33,600 of the population of Town X would vote no to building a park in the middle of the town.

D) Of the surveyed, 56 voted no to building a park in the middle of the town.

236. **Level:** Easy | **Skill/Knowledge:** Inference from sample statistics and margin of error | **Testing Point:** Calculating the sample standard of error

Key Explanation: Choice A is correct. Here p is the proportion. The standard error of p is given by $\sqrt{\left(\frac{p(1-p)}{n}\right)}$, where n is the number of employees interviewed.
Standard Error =
$$\sqrt{\left(\frac{0.9(1-0.9)}{1000}\right)}=\sqrt{\left(\frac{(0.9\times 0.1)}{1000}\right)}=0.0094\approx 0.009.$$

Distractor Explanations: Choice B is incorrect and mostly results from an error in calculating the standard error or applying the wrong formula for standard error. **Choice C** is incorrect and mostly results from an error in calculating the standard error or applying the wrong formula for standard error. **Choice D** is incorrect and mostly results from an error in calculating the standard error or applying the wrong formula for standard error.

237. **Level:** Easy | **Skill/Knowledge:** Inference from sample statistics and margin of error | **Testing Point:** Extrapolating from sample to population

Key Explanation: To solve the problem, set up a proportion as follows:

$$\frac{(defective\ sample\ widgets)}{(total\ sample\ widgets)}=\frac{(defective\ \ widgets)}{(total\ \ widgets)}$$

$$\frac{4}{300}=\frac{x}{12{,}000}.$$

Cross multiplying yields

$$300x = 4 \times 12{,}000.$$

Dividing both sides of the equation by 300 results in

$$x=\frac{48{,}000}{300}=160.$$

238. **Level:** Easy | **Skill/Knowledge:** Inference from sample statistics and margin of error | **Testing Point:** Understanding basic study design and interpreting sample proportion

Key Explanation: Choice A is correct since it is the proportion of the sample of 10,000 people.

Distractor Explanations: Choice B is incorrect and is the result of error in understanding the concept of sample and population proportion. **Choice C** is incorrect and is the result of error in understanding the concept of sample and population proportion. **Choice D** is incorrect and is the result of error in understanding the concept of sample and population proportion.

239. **Level:** Hard | **Skill/Knowledge:** Inference from sample statistics and margin of error | **Testing Point:** Margin of error

Key Explanation: Choice C is correct. This statement is true since the margin of error is 1.2 feet. Therefore the average height would lie between 34.6 ±1.2. This would mean that the average height lies between 33.4 feet to 35.8 feet.

Distractor Explanations: Choice A is incorrect because it is not certain. **Choice B** is incorrect because it is not certain. **Choice D** is incorrect because it is not certain.

240. **Level:** Medium | **Skill/Knowledge:** Inference from sample statistics and margin of error **Testing Point:** Calculating the margin of error

Key Explanation: The correct answer is 12.2. Use the formula $\frac{Maximum\ limit - Minimum\ limit}{2}$ to find the margin of error. Substituting the given data yields a Margin of error

$= \frac{160 - 135.6}{2} = \frac{24.4}{2} = 12.2.$

241. **Level:** Easy | **Skill/Knowledge:** Inference from sample statistics and margin of error | **Testing Point:** Making statistical inferences

Key Explanation: Choice A is correct. Increasing the sample size of the study would improve the accuracy of their study and thus reduce the margin of error.

Distractor Explanations: Choice B is incorrect. This option would increase the margin of error. **Choice C** is incorrect. This option would increase the margin of error. **Choice D** is incorrect. This would make the study biased.

242. **Level:** Medium | **Skill/Knowledge:** Inference from sample statistics and margin of error **Testing Point:** Extrapolating from sample data to a population

Key Explanation: Choice C is correct. The sample proportion of 17.8% is the percentage of households in the representative sample with cars. Multiplying this percentage to the total number of households in City X yields 0.178 × 21,000 = 3,738.

Distractor Explanations: Choice A is incorrect and may result from calculating half of the actual estimate. **Choice B** is incorrect is incorrect and may result from multiplying 10% instead of 17.8%. **Choice D** is incorrect and may result from calculating twice the actual estimate.

243. **Level:** Medium | **Skill/Knowledge:** Inference from sample statistics and margin of error **Testing Point:** Making inferences from sample data

Key Explanation: Choice C is correct. The study can only be generalized to elementary school teachers.

Distractor Explanations: Choice A is incorrect. This statement does not consider the education level. **Choice B** is incorrect. This statement does not consider the education level. **Choice D** is incorrect because substantial information is given to make a conclusion.

244. **Level:** Easy | **Skill/Knowledge:** Inference from sample statistics and margin of error | **Testing Point:** Making inferences from sample data

Key Explanation: Choice B is correct. From the survey, 56% of the sample group of residents in Town X voted yes to building the park. This would approximate to 56% of the town's population which is 60,000(56%) = 33,600. This is the estimated number of residents in Town X that would vote for building the park.

Distractor Explanations: Choice A is incorrect. This statement is false because 26,400 is the estimated number of people who are against building a park in the middle of the town. **Choice C** is incorrect. This statement is false because 56% of the sample group voted yes. **Choice D** is incorrect. This statement is false because 56 is not the number of residents but the percentage of residents that voted yes.

Evaluating statistical claims: observational studies and experiments

245

A study took a group of students and randomly divided them into two groups. The first group contained 1,000 students and the second 600 students. One group was told to wake up early in the morning and study, while the other group was told not to wake up early in the morning. Researchers then compared how each group got grades on the class test. Out of 1,000 students in the first group, 852 students improved their grades and in the second group, 48 improved their grades. What conclusion can be drawn from this experiment?

I. Most students who did not wake up early improved their grades.

II. Most students who did not wake up early have not improved their grades.

III. Most students who wake up early have improved their grades.

A) I alone is true.

B) I and II are true.

C) I and III are true.

D) II and III are true.

246

A random survey took a sample of 1,800 patients who were going for surgery and advised 1,000 of them to take a particular medicine for a week before the surgery day. For the rest 800, the medicine was not given. They also collected the data about the quick healing (healing before three days) of patients who took the medicine for a week. Out of the 1,000 patients who took the medicine, quick healing was observed in 808 patients, and out of the 800 patients who did not take the medicine, quick healing was observed only in 72 patients. What conclusion can be drawn from this experiment?

I. Most patients who did not take the medicine were healed quickly.

II. Most patients who did not take the medicine were not healed quickly.

III. Most patients who took the medicine were healed quickly.

A) I alone is true.

B) II alone is true.

C) I and III are true.

D) II and III are true.

247

A scientist wants to study how the amount of water humans drink affects their diets. Therefore, the scientist studies a group of people's daily water drinking usage. Which of the following is true?

A) This is an example of an observational study only.

B) This is an example of an experimental study only.

C) This is an example of both an observational study and an experimental study.

D) This is neither an example of an observational study nor an example of an experimental study.

248

A random survey took a sample of 1,600 patients who suffer from a particular disease. 900 of them are to take a particular medicine for two months. For the rest 700 the medicine was not given. They also collected the data after two months about the curing from disease of patients who took the medicine. Out of the 900 patients who took the medicine, curing from the disease was observed in 810 patients and out of the 700 patients who did not take the medicine the curing was observed only in 56 patients. What conclusion can be drawn from this experiment?

I. Most patients who did not take the medicine were cured.

II. Most patients who did not take the medicine were not cured.

III. Most patients who took the medicine were cured.

A) I alone is true.

B) II alone is true.

C) I and III are true.

D) II and III are true.

249

A fitness company has a running drink, which is consumed by 10 runners in the race. 7 of the 10 runners performed significantly better in their races compared to other runners who did not consume the drink. Which of the following statements can best be concluded about the running drink?

A) The drink helps all runners perform better in their races.

B) To perform better in races, runners should take the running drink.

C) The running drink cannot be linked to the runner's performance.

D) The running drink improves the running performance of some of the runners.

245. **Level:** Easy | **Skill/Knowledge:** Evaluating statistical claims: observational studies and experiments
Testing Point: Evaluating experimental studies

Key Explanation: Choice D is correct. II is true as only $\frac{48}{600}$ or 8% of the students who did not wake up early have improved. III is true as $\frac{852}{1,000}$ or 85.2% of the students improved their grades.

Distractor Explanations: I is not true since only $\frac{48}{600}$ or 8% improved their grades. **Choice A** is incorrect because it states that statement I is true. **Choice B** is incorrect because it states that statement I is true. **Choice C** is incorrect because it states that statement I is true.

246. **Level:** Easy | **Skill/Knowledge:** Evaluating statistical claims: observational studies and experiments
Testing Point: Making inferences from experiment data

Key Explanation: Choice D is correct. II is true as only $\frac{72}{800}$ or 9% of the patients who did not take the medicine have healed quickly. III is true as $\frac{808}{1,000}$ or 80.8% of the patients have healed quickly.

Distractor Explanations: I is not true since only $\frac{72}{800}$ or 9% healed quickly. **Choice A** is incorrect because it states that statement I is true. **Choice B** is incorrect because statement III is also true. **Choice C** is incorrect because it states that statement I is true.

247. **Level:** Easy | **Skill/Knowledge:** Evaluating statistical claims: observational studies and experiments
Testing Point: Determining the difference between observational studies and experiments

Key Explanation: Choice A is the correct answer. In an observational study, researchers study how participants perform certain behaviors or activities without telling them what methods or behaviors to choose.

Distractor Explanations: Choice B is incorrect. In experimental studies, researchers introduce an intervention and study the effects. **Choice C** is incorrect. In experimental studies, researchers introduce an intervention and study the effects. **Choice D** is incorrect because this is an example of observational study.

248. **Level:** Easy | **Skill/Knowledge:** Evaluating statistical claims: observational studies and experiments
Testing Point: Interpreting sample data results

Key Explanation: Choice D is correct. II is true as only $\frac{56}{700}$ or 8% of the patients who did not take the medicine got cured, rest 92% not cured. III is true as $\frac{810}{900}$ or 90% of the patients who took the medicine got cured. Hence Statements II and III are true.

Distractor Explanations: I is not true since only $\frac{72}{800}$ or 9% who did not take the medicine got cured. **Choice A** is incorrect because statement I is not true. **Choice B** is incorrect because statement III is also true. **Choice C** is incorrect because statement I is not true.

249. **Level:** Hard | **Skill/Knowledge:** Evaluating statistical claims: observational studies and experiments
Testing Point: Evaluating experimental study data

Key Explanation: Choice D is correct. Since only 7 out of 10 runners performed better after consuming the drink, the drink did not enhance the performance of all of them. Therefore, some but not all of the runners noticed a significant improvement in their races after consuming the running drink.

Distractor Explanations: Choice A is incorrect. Not all runners performed better after consuming the running drink. **Choice B** is incorrect. This implies a cause-and-effect relationship between the running drink and the runner's performance. **Choice C** is incorrect. The result on the 7 runners concludes that the running drink affects the runner's performance.

Chapter 6

Geometry and Trigonometry

This chapter includes questions on the following topics:

- Area and volume
- Lines, angles, and triangles
- Right triangles and trigonometry
- Circles

250

Jen is choosing new triangular tiles for her bathroom floor. She has two options as shown below. The two tiles, triangle *ABC* and triangle *abc*, are similar triangles. If Jen needs enough tiles to cover 540 square inches of the floor in her bathroom, how many more tiles of triangle *abc* would she need to purchase if she chose that option over triangle *ABC*?

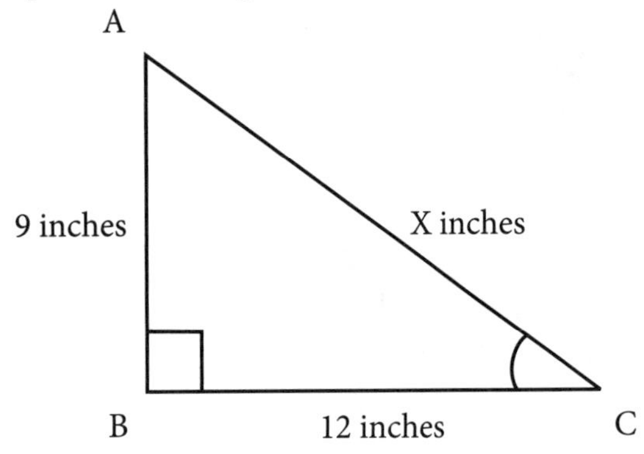

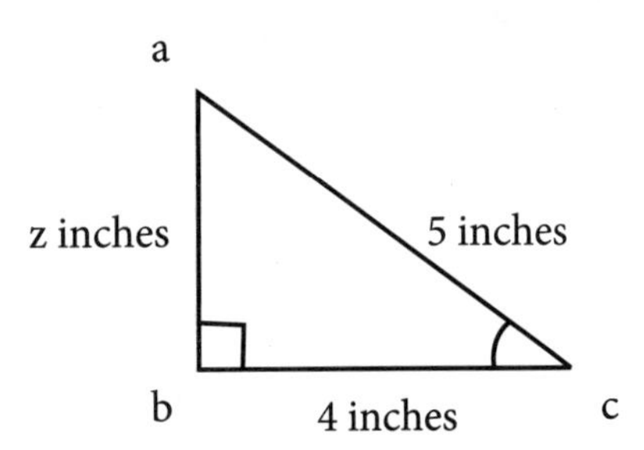

(Note: Figure is not drawn to scale.)

251

Cindy is choosing a piñata for a birthday party. The piñata she wants to buy is a sphere with a radius of 7 inches. If she wants to fill the piñata with 1,400 in^3 of candy, how much empty space will there be inside the piñata, in in^3? Round to the nearest integer. (Take $\pi = \frac{22}{7}$)

252

Eliza is moving to a new house. If her living room has an area of 175 square feet and she has a couch and table that measure 10 × 7 feet in total, how much empty space will she have left in her living room once she arranges the couch and table?

A) 70 ft^2

B) 105 ft^2

C) 150 ft^2

D) 175 ft^2

253

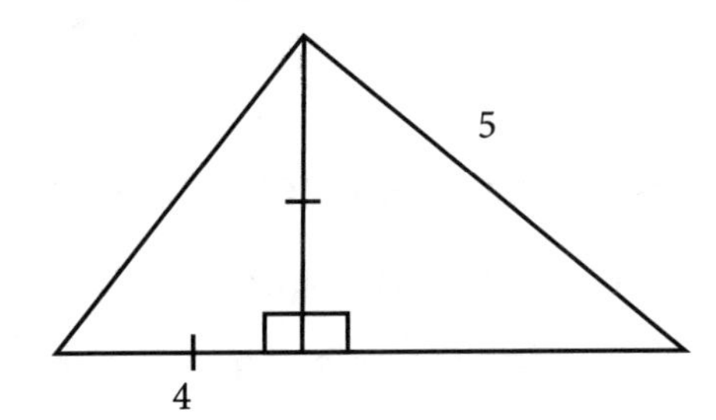

(Note: Figure is not drawn to scale.)

What is the total area of the figure above?

A) 6

B) 8

C) 12

D) 14

254

The width of a rectangular bulletin board being decorated for a football game is (a) meters, and the length (b) is 3 meters shorter than twice the width. Which of the following expressions represent the area of the bulletin board?

A) $a(2a + 3)$

B) $a(2a - 3)$

C) $2a^2 - 3$

D) $4a + 3$

255

John wants to put a new carpet in his movie room. The movie room is 6.25 yards long and 4.37 yards wide. The carpet costs $9.75 per square foot. What will John pay to put the new carpet in his movie room? (1 yard = 3 feet)

A) $27.35

B) $245.81

C) $266.30

D) $2,396.67

256

If the circumference of a circular region equals exactly 6π inches, which is the area of the region?

A) 3π

B) 9π

C) 12π

D) 36π

257

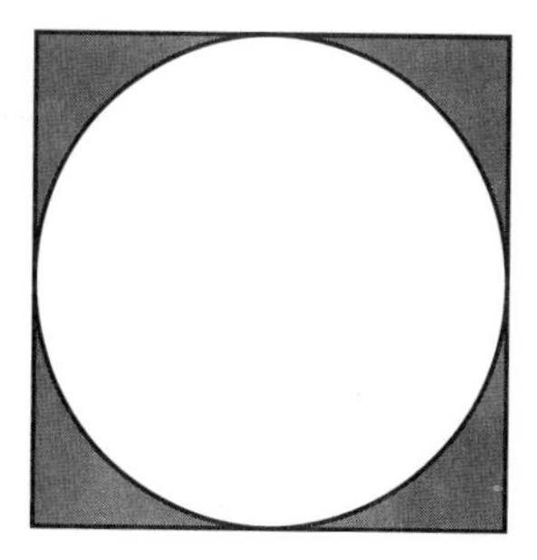

In the figure above, the circle is inscribed within the square. If the diameter of the circle is 12, what is the area of the shaded region?

A) 144

B) 36π

C) $144 - 36\pi$

D) $36 - 9\pi$

258

The volume of a particular cube is 27 cubic inches. What is the area of the base of the cube in square inches?

259

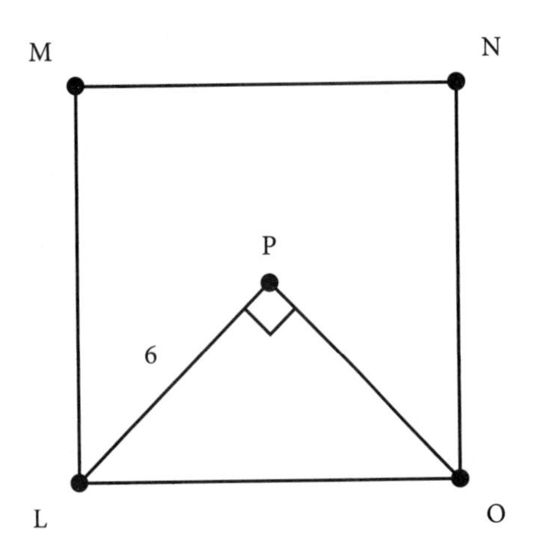

In the figure above, $LMNO$ is a square, and triangle LPO is an isosceles triangle. If $LP = 6$, what is the area of the square?

A) 24

B) 36

C) 48

D) 72

260

If a 440m track encompasses a circular soccer field, what is the area of half the soccer field? (Use π as $\frac{22}{7}$)

A) 140 m^2

B) 4,900 m^2

C) 7,700 m^2

D) 15,400 m^2

261

The radius of a sphere is equal to the length of a cube. If the cube has a surface area of 54 cm^2, what is the volume of the sphere in terms of π?

A) 9π

B) 18π

C) 27π

D) 36π

262

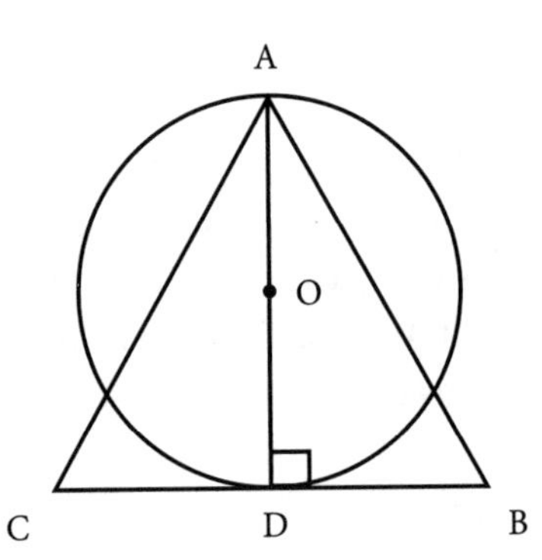

In the figure above, the area of the circle with center O is 81π square units. What is the area of equilateral triangle ABC?

A) $108\sqrt{3}$ square units

B) $116\sqrt{3}$ square units

C) $124\sqrt{3}$ square units

D) $142\sqrt{3}$ square units

263

Paul has a ball with a volume of 36π. Marc has a bigger ball of the same material. If the radius of Marc's ball is twice the radius of Paul's ball, what is the volume of Marc's ball?

A) 64π

B) 128π

C) 144π

D) 288π

264

A large circular frisbee has a diameter of 2 feet and a height of 3 inches. If Jennifer stacks 15 frisbees on top of each other to prepare for the field day, what is the volume of the stack in cubic feet?

A) $\frac{\pi}{4}$

B) $\frac{15\pi}{4}$

C) 6π

D) 45π

250. **Level:** Medium | **Skill/Knowledge:** Area and volume | **Testing Point:** Finding areas of similar triangles using scale factor

Key Explanation: To find the difference between the number of tiles needed to cover 540 square inches, first find the area of each triangular tile. The area of a triangle can be found by the formula $A = \frac{1}{2}bh$ where $b =$ base and $h =$ height. Solving the area of triangle ABC yields $\frac{1}{2}(12)(9)$ or 54 square inches. To find the area of triangle abc, first, determine the value of z. Using the properties of similar triangles to determine the value of z yields the ratio of the sides $\frac{12}{4} = \frac{9}{z}$. Simplifying the equation yields $3 = \frac{9}{z}$. Multiplying z and dividing 3 from both sides of the equation yields $z = \frac{9}{3}$ or $z = 3$. Solving for the area of triangle abc yields $\frac{1}{2}(4)(3)$ or 6 square inches. To get the number of tiles required for each option, divide the area of the floor by the area of the corresponding tile. Solving for the number of triangle ABC tiles needed yields $\frac{540}{54}$ or 10 tiles. Solving for the number of triangle abc tiles needed yields $\frac{540}{6}$ or 90 tiles. Calculating the difference between the number of tiles needed yields $90 - 10 = 80$.

251. **Level:** Easy | **Skill/Knowledge:** Area and volume **Testing Point:** Finding the volume of a sphere

Key Explanation: The correct answer is 37. The formula to find the volume of a sphere is $V = \frac{4}{3}\pi r^3$. Substituting the given data to find the volume of the piñata yields

$$V = \left(\frac{4}{3}\right)\pi r^3 = \left(\frac{4}{3}\right)\pi 7^3 = \left(\frac{4}{3}\right)\left(\frac{22}{7}\right)(7)(7)(7)$$

$= 1{,}437.33$.

Solving the volume of the empty space inside the piñata yields $1{,}437.33 - 1{,}400 = 37.33$. Rounding the answer to the nearest integer yields 37.

252. **Level:** Easy | **Skill/Knowledge:** Area and volume **Testing Point:** Using definitions, properties, and theorems relating to area and volume to solve problems

Key Explanation: Choice B is correct. To find the space left in the room, find the difference between the area of the living room and the area that will be occupied by the furniture. The area that will be occupied by the furniture is $10 \times 7 = 70\ ft^2$. Then, subtract this value from the area of the entire living room which yields $175 - 70 = 105\ ft^2$.

Distractor Explanations: Choice A is incorrect and may result from calculating only the area that will be occupied by the furniture. **Choice C** is incorrect and may result from a conceptual or calculation error. **Choice D** is incorrect and may result from a conceptual or calculation error.

253. **Level:** Medium | **Skill/Knowledge:** Area and volume | **Testing Point:** Finding the area of composite polygons

Key Explanation: Choice D is correct. To solve for the area of the whole triangle, determine the areas of the left and right triangles individually and then add them together. Since the triangle on the left has two equal interior angles and a right angle, then it is a 45–45–90 angle. This means that the length of the legs is congruent. So, the altitude is also equal to 4. Solving for the area of the left triangle yields $\frac{1}{2}(4)(4) = 8$. Since the other triangle has a hypotenuse of 5 and a leg of 4, then it is a 3–4–5 triangle. So, the length of the base must be 3. The area of this triangle is $\frac{1}{2}(4)(3)=6$. The combined area is 6 + 8 = 14.

Distractor Explanations: Choice A is incorrect and may result from solving only the area of the right triangle. **Choice B** is incorrect and may result from solving only the area of the left triangle. **Choice C** is incorrect and may result from assuming that both left and right triangles are a 3–4–5 special triangle.

254. **Level:** Easy | **Skill/Knowledge:** Area and volume **Testing Point:** Converting English into Algebra to solve area problem

Key Explanation: Choice B is correct. The formula to find the area of a rectangle is *length* × *width*. Since $a = width$ and $b = length$, the area of the rectangle is a times b. However, since the length b is 3 meters shorter than twice the width, b can be expressed as $2a - 3$. Substituting $2a - 3$ for b in the area equation yields
Area = $a \times b = a \times (2a - 3) = a(2a - 3)$ or **Choice B**.

Distractor Explanations: Choice A is incorrect and is the result of error in converting the English into Algebra or calculating the wrong area of a rectangle. **Choice C** is incorrect and is the result of error in converting the English into Algebra or calculating the wrong area of a rectangle. **Choice D** is incorrect and is the result of error in converting the English into Algebra or calculating the wrong area of a rectangle.

255. **Level:** Easy | **Skill/Knowledge:** Area and volume **Testing Point:** Finding area rectangle and units conversion

Key Explanation: Choice D is correct. The first step is to convert yards to feet for the length of the rectangle. To do this, multiply each dimension by 3; *length* = 6.25 *yds* × 3 = 18.75 *ft*, *width* = 4.37 *yds* × 3 = 13.11 *ft*. Find the area of the rectangle by using the formula A = length × width, or 18.75 × 13.11 = 245.8125 ft^2. Find the cost to put in the carpet by multiplying the area by the cost per square foot, 245.8125 ft^2 × \$9.75 = \$2,396.671875. This value rounds to \$2,396.67 or **Choice D**.

Distractor Explanations: Choice A is incorrect and is likely the result of arithmetic error in converting units. **Choice B** is incorrect and likely the result of arithmetic errors in converting units. **Choice C** is incorrect and is likely the result of arithmetic error in converting units.

256. **Level:** Easy | **Skill/Knowledge:** Area and volume **Testing Point:** Using circumference of a circle to find its area

Key Explanation: Choice B is correct. If the circumference is exactly 6π inches, find the radius using the formula of circumference which is $C = 2\pi r$ where C is the circumference and r is the radius. Substituting 6π to the formula yields $6\pi = 2\pi r$. Dividing both sides of the equation by 2π yields $3 = r$ or $r = 3$. To find the Area, use the formula $A = \pi r^2$. Substituting the radius yields $A = \pi(3)^2 = 9\pi$.

Distractor Explanations: Choice A is incorrect and reflects error in calculating the circumference and area of a circle. **Choice C** is incorrect and reflects error in calculating the circumference and area of a circle. **Choice D** is incorrect and reflects error in calculating the circumference and area of a circle.

257. **Level:** Hard | **Skill/Knowledge:** Area and volume **Testing Point:** Finding the area of shaded region with inscribed shape

Key Explanation: Choice C is correct. Find the area of the circle by using the formula $A = \pi r^2$. Since the diameter is 12, then the radius is $\frac{12}{2} = 6$. Substituting the radius to the formula yields $A = \pi(6^2) = 36\pi$. Next, find the area of the square by using the formula $A = s^2$. Since the circle is inscribed in the square, the diameter of the circle is equal to the side length of the square. This means that $s = 12$. Substituting the length of the side to the formula yields $A = 12^2 = 144$. Now subtract the area of the circle from the area of the square which yields $144 - 36\pi$.

Distractor Explanations: Choice A is incorrect and may result from solving the whole area of the square. **Choice B** is incorrect and may result from solving the area of the circle. **Choice D** is incorrect and may result from using 6 as the diameter of the circle and the side length of the square.

258. **Level:** Easy | **Skill/Knowledge:** Area and volume
Testing Point: Finding the area of a square prism

Key Explanation: The correct answer is 9. To find the area of the base, find the length of one of the edges of the cube. Since the volume is $s^3 = 27$, it follows that $s = \sqrt[3]{27} = 3$. Since the length of one of the edges of the cube is 3, we use the formula for the area of a square. Therefore, the area of the base of the cube is $A = s^2 = (3)^2 = 9$.

259. **Level:** Easy | **Skill/Knowledge:** Area and volume
Testing Point: Using theorems relating to squares and isosceles triangles to solve problems

Key Explanation: Choice D is correct. In the figure, $LMNO$ is a square, and triangle LPO is an isosceles. If $LP = 6$, then $PO = 6$. Using the Pythagorean theorem to find the length of LO yields $6^2 + 6^2 = LO^2$. Simplifying the equation yields $72 = LO^2$. Getting the square root of both sides yields $\sqrt{72} = LO$. Since LO is a side of the square $LMNO$, the area of the square is $(\sqrt{72})^2$ or 72.

Distractor Explanations: Choice A is incorrect resulting from conceptual or arithmetic errors. **Choice B** is incorrect resulting from conceptual or arithmetic errors. **Choice C** is incorrect resulting from conceptual or arithmetic errors.

260. **Level:** Hard | **Skill/Knowledge:** Area and volume
Testing Point: Finding the area of a circle given its circumference

Key Explanation: Choice C is correct. The length of the track represents the circumference of the soccer field. Getting the diameter of the circular field yields $440\ m = \pi D$ or $\frac{22}{7}D = 440$. Multiplying 7 and dividing 22 from both sides of the equation yields $D = 140$. Hence, the radius of the soccer field is $\frac{140}{2}\ m$ or $70\ m$. The area of half of the circular field would therefore be $\frac{1}{2} \times \frac{22}{7} \cdot 70^2$ or $7{,}700\ m^2$.

Distractor Explanations: Choice A is incorrect. This is the diameter of the soccer field. **Choice B** is incorrect and is a result of calculation or conceptual error. **Choice D** is incorrect and is the area of the whole circle.

261. **Level:** Easy | **Skill/Knowledge:** Area and volume
Testing Point: Finding the volume of cube given surface area

Key Explanation: Choice D is correct. The surface area of a cube is given by the formula $6a^2$, given that a is one side of the cube. Since the surface area of the cube is 54, $6a^2 = 54$. Dividing both sides of the equation by 6 yields $a^2 = 9$. Getting the square root of both sides of the equation yields $a = 3$. The radius of the sphere will be 3 cm. The volume of a sphere is given by the formula $\frac{4}{3}\pi r^3$ which results to $\frac{4}{3}\pi(3)^3 = 36\pi$.

Distractor Explanations: Choice A is incorrect and may be a result of a conceptual error or miscalculation. **Choice B** is incorrect and may be a result of a conceptual error or miscalculation. **Choice C** is incorrect and may be a result of a conceptual error or miscalculation.

262. **Level:** Easy | **Skill/Knowledge:** Area and volume **Testing Point:** Using area of circle and concepts of equilateral triangles

Key Explanation: Choice A is correct.
Area of the circle = πr^2.
If r is the radius, $\pi r^2 = 81\pi$ which yields $r = 9$.
Diameter $AD = 18$.
Triangle ABC is equilateral.
Angle $ADB = 90°$.
Hence AD becomes altitude of ABC.
If a is the side of the equilateral triangle then altitude $= \frac{\sqrt{3}}{2}a$ [formula].

So: $\frac{\sqrt{3}}{2}a = 18$

$a = \frac{36}{\sqrt{3}}$

$a^2 = \frac{1{,}296}{3} = 432.$

By formula, area of the equilateral triangle is $\frac{\sqrt{3}}{4}a^2$ square units.

So area $= \frac{432\sqrt{3}}{4} = 108\sqrt{3}$ square units.

Hence **Choice A** is correct.

Distractor Explanations: Choice B is incorrect because when we calculate the area, they do not equal to $108\sqrt{3}$. **Choice C** is incorrect because when we calculate the area, they do not equal to $108\sqrt{3}$. **Choice D** is incorrect because when we calculate the area, they do not equal to $108\sqrt{3}$.

263. **Level:** Medium | **Skill/Knowledge:** Area and volume | **Testing Point:** Working with similar geometric three dimensional shapes

Key Explanation: Choice D is correct. The ratio of the radius of Paul's ball to Marc's ball is 1:2. Therefore, the ratio of their volumes would be $1^3:2^3$ or 1:8. Since Paul's ball has a volume of 36π, the ratio will be $\frac{Volume\ of\ Marc's\ ball}{36\pi} = \frac{8}{1}$.
Multiplying both sides of the equation by 36π yields Volume of Marc's ball = 288π.

Distractor Explanations: Choice A is incorrect. This is the value when you use the ratio of the radius to find the volume of Marc's ball. **Choice B** is incorrect and may result from a conceptual or calculation error. **Choice C** is incorrect and may result from a conceptual or calculation error.

264. **Level:** Easy | **Skill/Knowledge:** Area and volume **Testing Point:** Finding the volume of a cylinder

Key Explanation: Choice B is correct. First, find the volume of one frisbee by using the formula *Volume* = $\pi r^2 h$. The diameter is 2 *ft*, therefore, r = 1 *ft*. The height is 3 inches; converting it to feet yields $h = \frac{3in}{12}ft$ or $h = \frac{1}{4}ft$. Hence, the volume of one frisbee is $\frac{\pi 1^2 1}{4} = \frac{\pi}{4}$. Therefore, the volume of 15 frisbees is $\frac{15\pi}{4}$.

Distractor Explanations: Choice A is incorrect and may result from solving the volume of 1 frisbee. **Choice C** is incorrect and may result from solving the volume of 24 frisbees. **Choice D** is incorrect and may result from solving the volume of 180 frisbees.

265

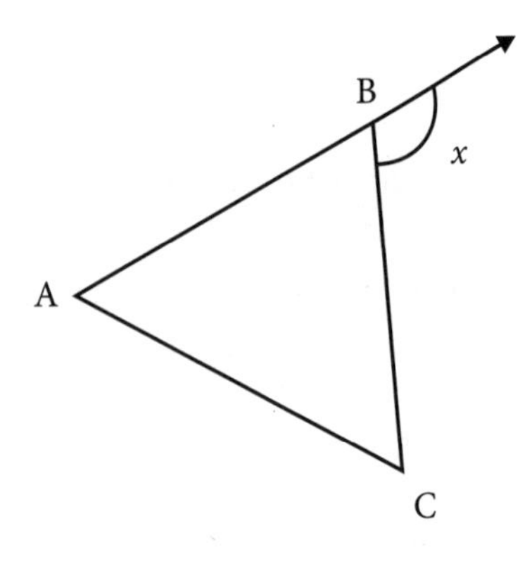

(Note: Image is not drawn to scale.)

If $\angle A = 62°$ and $\angle C = 51°$, what is the value of $\angle x$?

A) 67°

B) 113°

C) 118°

D) Cannot be determined

266

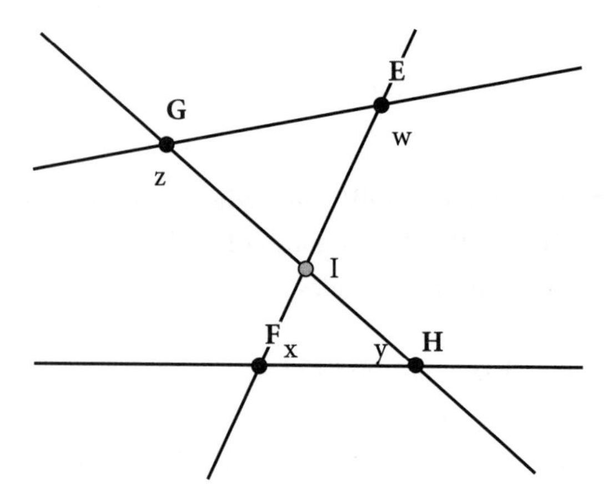

In the above figure, if $x + y = 108°$, then what is the value of $z + w$ (in degrees)?

A) 72

B) 108

C) 252

D) 282

267

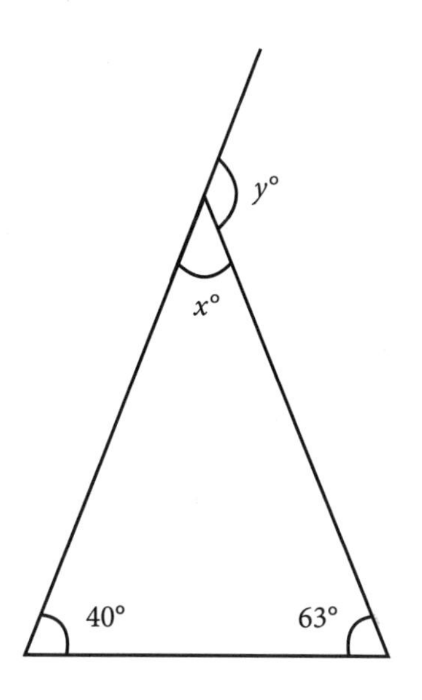

(Note: Figure is not drawn to scale.)

In the figure above, what is the value of y°?

A) 13°

B) 77°

C) 103°

D) 117°

268

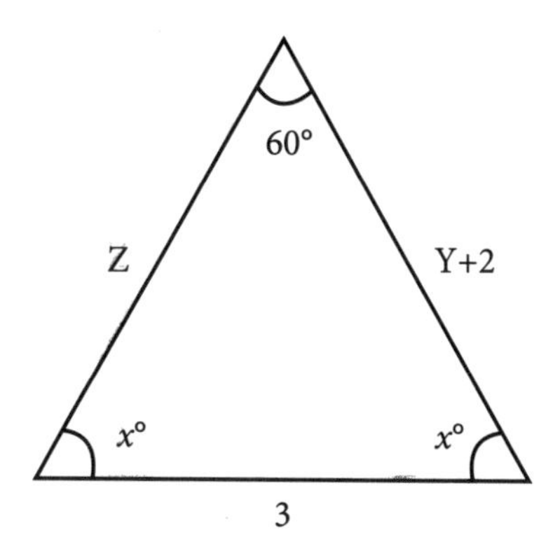

According to the triangle above, what is the value of side Z?

269

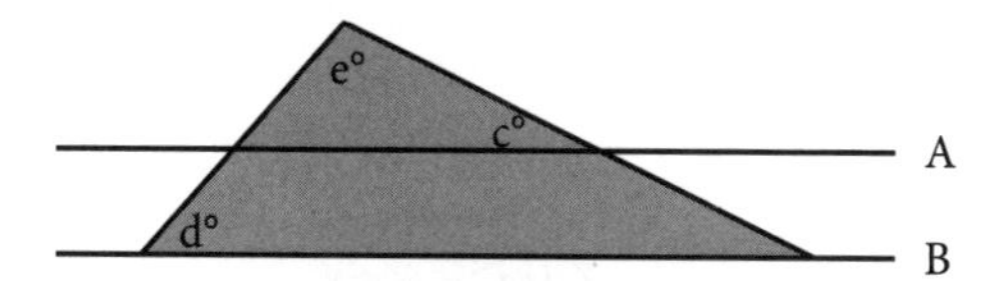

In the figure above, lines A and B are parallel, $c = 35$, and $d = 55$. Which best describes the shaded triangle type?
(Note: Figure is not drawn to scale.)

A) Isosceles Right Triangle

B) Scalene Right Triangle

C) Equilateral Triangle

D) Scalene Obtuse Triangle

270

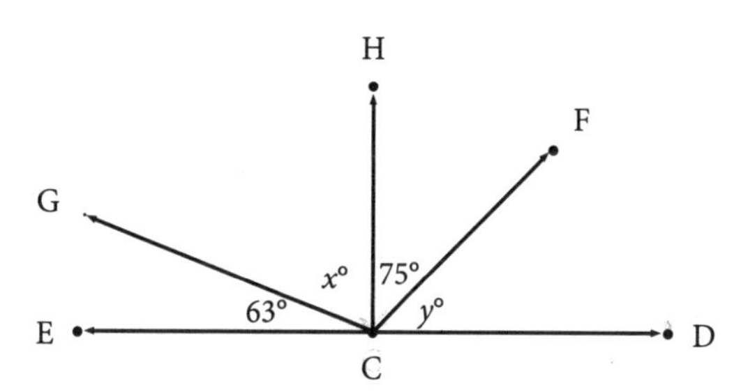

In the figure above, assuming E, C, and D are collinear, what is the sum of $x + y$ (in degrees)?

271

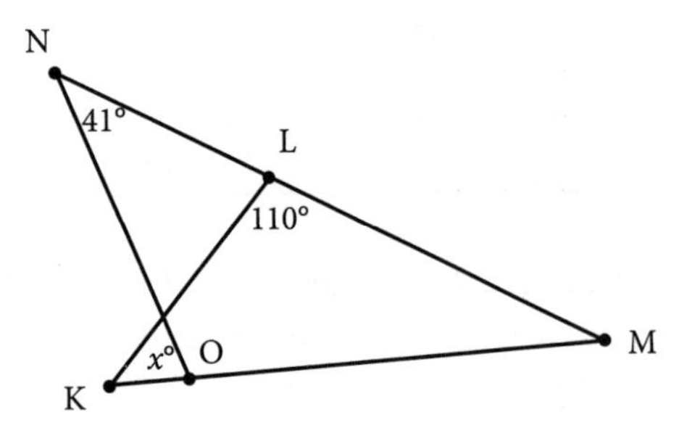

In the figure above, the length of KL = the length of ML. What is the value of x ($\angle NOK$) in degrees?

A) 35

B) 70

C) 76

D) 104

272

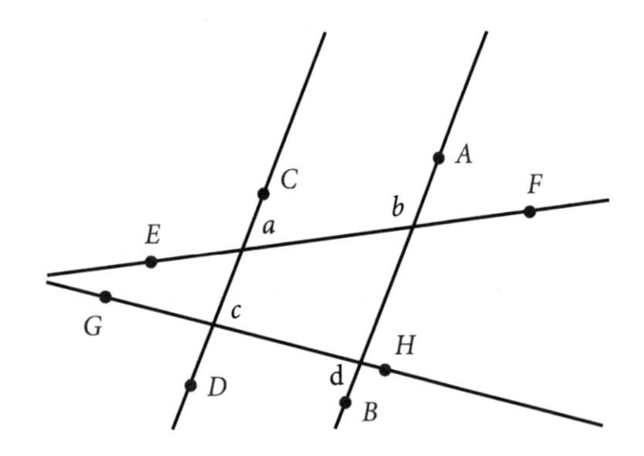

In the above figure AB and CD are parallel lines. EF and GH are two transversals.
What is $b + d$ in terms of a and c?

A) $180 + a + c$

B) $180 - a - c$

C) $180 - a + c$

D) $180 + 2(a + c)$

273

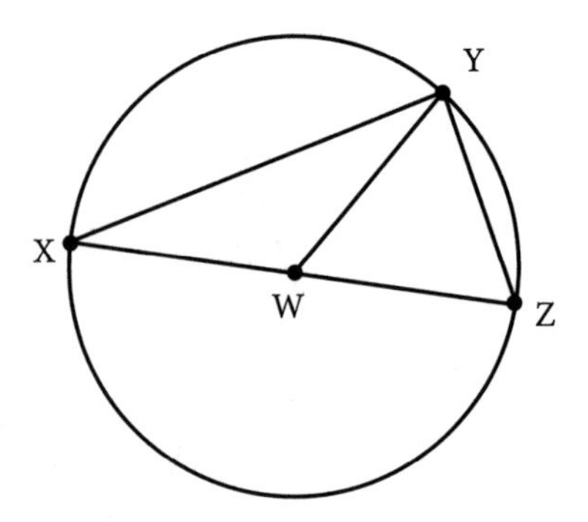

In the figure above, triangle XYZ is inscribed in the circle with center W and diameter XZ. If $YZ = YW$, how many degrees is angle XWY?

274

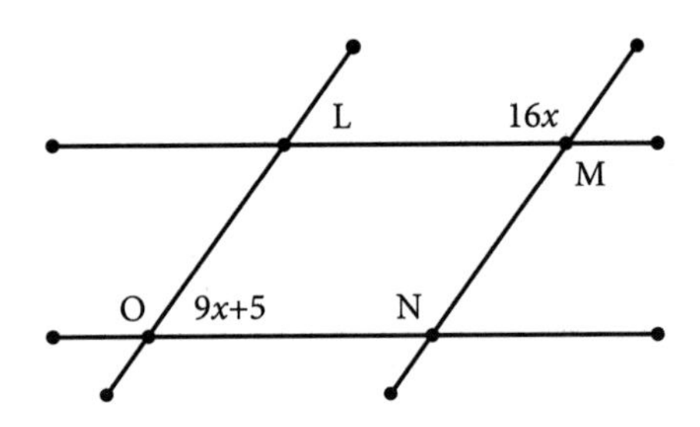

If $LMNO$ is a parallelogram, what is the difference between $\angle MNO$ and $\angle LMN$?

A) 7°

B) 44°

C) 68°

D) 70°

275

The measure of angle P is $\frac{7}{18}\pi$ radians. What is the measure of angle P in degrees?

276

If line m // line n, what is the value of angle z?

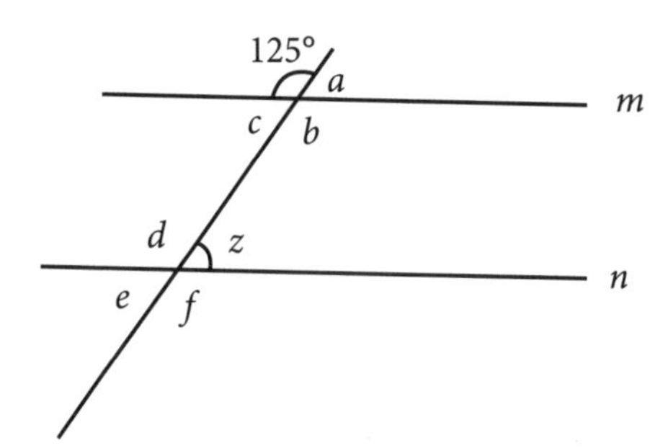

277

Point A has coordinates of (6, 4) and Point B has coordinates of (0, –8). Point C is the midpoint of segment AB. Which one of the following lines does Point C lie on?

A) $y = x + 4$

B) $y = x + 3$

C) $y = x - 5$

D) $y = x - 2$

265. **Level:** Easy | **Skill/Knowledge:** Lines, angles, and triangles | **Testing Point:** Using concepts of supplementary angles and angle sum property of triangles to solve problems

Key Explanation: Choice B is correct. The exterior angle theorem states that the measure of each exterior angle of a triangle is equal to the sum of the opposite and non-adjacent interior angles. So, the value of $\angle x$ is the sum of $\angle A$ and $\angle C$. Therefore, the value of $\angle x$ is 51° + 62° = 113°.

Distractor Explanations: Choice A is incorrect and may result from calculating the value of $\angle B$. **Choice C** is incorrect and may result from calculating the sum of $\angle B$ and $\angle C$. **Choice D** is incorrect because all the needed data can be obtained from the problem.

266. **Level:** Easy | **Skill/Knowledge:** Lines, angles, and triangles | **Testing Point:** Using concepts of supplementary angles and angle sum property of triangles to solve problems

Key Explanation: Choice C is correct. Using the given equation $x + y = 108°$ find the value of $\angle FIH$. Applying the Angle sum property of a triangle in ΔFIH yields $x + y + \angle FIH = 180°$.

Substituting the value of $x + y$ yields 108°+ $\angle FIH$ =180°. Subtracting 108 from both sides of the equation yields $\angle FIH$ =180° – 108° = 72°. Since $\angle FIH$ and $\angle GIE$ are vertically opposite angles then $\angle FIH = \angle GIE$.

Therefore, $\angle GIE$ =72°.

Solving for the angles in ΔGEI yields $\angle EGI$ = 180° – z and $\angle GEI$ =180° – w.

Using the Angle sum property of a triangle in ΔGEI yields $\angle GIE + \angle EGI + \angle GEI$ =180°.

Substituting the values of $\angle GIE$, $\angle EGI$ and $\angle GEI$ yields 72° + 180° – z + 180° – w = 180°.

Combining like terms and simplifying the equation yields 252° = $z + w$.

Distractor Explanations: Choice A is incorrect and may result from solving for the value of $\angle GIE$. **Choice B** is incorrect and may result from solving for the sum of $\angle EGI$ and $\angle GEI$. **Choice D** is incorrect and may result from a conceptual or calculation error.

267. **Level:** Easy | **Skill/Knowledge:** Lines, angles, and triangles | **Testing Point:** Using concepts of supplementary angles and angle sum property of triangles to solve problems

Key Explanation: Choice C is correct. The sum of the interior angles of a triangle is 180°. Therefore, the sum of the interior angles in the figure can be represented by x + 40° + 63° = 180. Subtracting 40 and 63 to both sides of the equation yields x = 77. Angles x and y are supplementary angles, which also have a sum of 180°. Therefore, y = 180° – 77° = 103°.

Distractor Explanations: Choice A is incorrect and may result from calculating the complementary angle of x. **Choice B** is incorrect because angles x and y cannot be equal. **Choice D** is incorrect and may result from calculating the sum of x and 40°.

268. **Level:** Easy | **Skill/Knowledge:** Lines, angles, and triangles | **Testing Point:** Using properties of equilateral triangles to solve problems

Key Explanation: The correct answer is 3. Determine the type of the triangle by solving for the values of the interior angles. Applying the Angle Sum Theorem yields 60° + x + x = 180°. Combining like terms and subtracting 60° from both sides of the equation yields $2x = 120°$. Dividing 2 from both sides of the equation yields $x = 60°$. Since all the interior angles are the same, the triangle is an equilateral triangle. Since an equilateral triangle also has equal sides, $Z = 3$.

269. **Level:** Medium | **Skill/Knowledge:** Lines, angles, and triangles | **Testing Point:** Using properties of triangles to solve triangles and determine triangle type

Key Explanation: Choice B is correct. If the $\angle c = 35°$ then the measure of the angle on the bottom right of the big triangle is the same as the measure of angle c, or 35 degrees since lines A and B are parallel and these two angles are corresponding angles and thus are equal. Since a triangle has 180 degrees, the measure of angle e is (180 – 35 – 55) or 90°. Since $\angle e$ measures 90 degrees, it is a right angle. Since all of the angles have different values, the triangle is a scalene right triangle.

Distractor Explanations: Choice A is incorrect and assumes incorrectly that two sides of the triangle are congruent. **Choice C** is incorrect because it does not take into account the right angle. **Choice D** is incorrect because it does not take into account the right angle.

270. **Level:** Easy | **Skill/Knowledge:** Lines, angles, and triangles | **Testing Point:** Using concepts of supplementary angles and linear pair

Key Explanation: The correct answer is 42. A straight line has 180 degrees. Therefore:

$63 + x + 75 + y = 180$

$x + y + 138 = 180$

$x + y = 42$.

271. **Level:** Medium | **Skill/Knowledge:** Lines, angles, and triangles | **Testing Point:** Using concepts of supplementary angles, isosceles triangles, and angle sum property of triangles to solve problems

Key Explanation: Choice C is correct. First, find the values of $\angle M$ & $\angle K$. Since ΔKLM is an isosceles triangle, $\angle K$ & $\angle M$ are congruent to each other. Since the sum of the inside angles in a triangle is 180°, then $180 - 110 = \angle K + \angle M$. Simplifying the equation yields $70 = \angle K + \angle M$. Since both angles are equal, $\angle K = \angle M = 70 \div 2 = 35$. Hence, $\angle K$ & $\angle M$ are both 35°. To find $\angle NOM$, find the sum of $\angle M$ and $\angle N$ and subtract the sum from 180. This yields 180 – (41 +35) = 104°. Since $\angle x$ is the supplement of $\angle NOM$, then $\angle x$ = 180 – 104 = 76.

Distractor Explanations: Choice A is incorrect and may result from solving the value of $\angle K$ and $\angle M$. **Choice B** is incorrect and may result from solving the value of $\angle NLK$. **Choice D** is incorrect and may result from solving the value of $\angle NOM$.

272. **Level:** Medium | **Skill/ Knowledge:** Lines, angles, and triangles | **Testing Point:** Using concepts of supplementary angles and parallel lines to solve problems

Key Explanation: Choice C is the correct answer. Given that $AB \parallel CD$ and EF is a transversal, then $a + b = 180°$ since a and b are co-interior angles.
Hence, $b = 180 - a$.
Given that GH is a transversal for AB and CD, then $d = c$ since c and d are alternate interior

angles.

Therefore, $b + d = 180 - a + c$.

Distractor Explanations: Choice A is incorrect and most likely results from error in calculating the value of b and d. **Choice B** is incorrect and most likely results from error in calculating the value of b and d. **Choice D** is incorrect and most likely results from error in calculating the value of b and d.

273. **Level:** Medium | **Skill/Knowledge:** Lines, angles, and triangles | **Testing Point:** Using concepts of supplementary angles, equilateral triangles, and angle sum property of triangles to solve problems

Key Explanation: 120° is the correct answer. Since the figure above shows that triangle XYZ is inscribed in the circle with center W and diameter XZ, then $YW = ZW$ because they are both radii. Since $YZ = YW = ZW$, then triangle WYZ is equilateral which means all angles are congruent. Since the sum of the inside angles of a triangle is 180°, each angle is 60°. Hence, $\angle ZWY = 60°$. Since $\angle XWY$ is the supplement of $\angle ZWY$, then $\angle XWY = 180 - 60 = 120°$.

274. **Level:** Easy | **Skill/Knowledge:** Lines, angles, and triangles | **Testing Point:** Using parallel line theorems to solve problems

Key Explanation: Choice B is correct. Since $LMNO$ is a parallelogram, we know that there are two sets of parallel lines being cut by two parallel transversals. Based on the Corresponding Angles theorem, $\angle MNO = 16x$. Since $\angle LON$ and $\angle LMN$ are opposite angles of a parallelogram, then $\angle LON = \angle LMN = 9x + 5$. Since $\angle MNO$ is a supplement to $\angle LMN$, then $16x + 9x + 5 = 180$ or $25x + 5 = 180$. Subtracting 5 from both sides of the equation yields $25x = 175$. Dividing both sides of the equation by 25 yields $x = 7$. It follows that $\angle MNO = 16x = 16(7) = 112$ and $\angle LMN = 9x + 5 = 9(7) + 5 = 63 + 5 = 68$. Therefore the difference between the two is $112 - 68 = 44$.

Distractor Explanations: Choice A is incorrect and reflects error in interpreting angle relationships. **Choice C** is incorrect and reflects error in interpreting angle relationships. **Choice D** is incorrect and reflects error in interpreting angle relationships.

275. **Level:** Easy | **Skill/Knowledge:** Lines, angles, and triangles | **Testing Point:** Converting from radians to degrees

Key Explanation: 70° is the correct answer. To convert radians to degrees, use the formula $\frac{180}{\pi} \times \theta$. Substituting the given angle yields $\frac{180}{\pi} \cdot \frac{7}{18}\pi$ or 70°.

276. **Level:** Easy | **Skill/Knowledge:** Lines, angles, and triangles | **Testing Point:** Finding the measure of angles in parallel lines

Key Explanation: 55° is the correct answer. Angles on a straight line add up to 180°. Therefore, $\angle c = 180 - 125 = 55°$. Since $\angle c$ and $\angle z$ are alternate interior angles, they are equal. Therefore, $z = 55°$.

277. **Level:** Medium | **Skill/Knowledge:** Lines, angles, and triangles | **Testing Point:** Calculating midpoint of a line segment and knowledge of graphs

Key Explanation: Choice C is the correct

answer. The coordinates of the midpoint of a line segment is calculated by taking the average of the x coordinates and the average of the y coordinates of the two points on the line segment. Thus, the midpoint of segment AB is given by $\left(\frac{6+0}{2}, \frac{4-8}{2}\right)$ or (3, –2). To determine the correct equation the midpoint coordinates lie on, substitute (3, –2) into each of the answer choices for the variables x and y and determine which answer choice makes the equation true. The only answer choice that makes the equation true is **Choice C** as $-2 = 3 - 5$ or $-2 = -2$.

Therefore, the midpoint lies on the line $y = x - 5$.

Distractor Explanations: Choice A is incorrect and is the result of error in calculating the midpoint of a line segment or in substituting points into equation. **Choice B** is incorrect and is the result of error in calculating the midpoint of a line segment or in substituting points into equation. **Choice D** is incorrect and is the result of error in calculating the midpoint of a line segment or in substituting points into equation.

278

In right triangle *ABC*, if $\angle B = 90°$ and $\cos A = \frac{1}{2}$, then what is *sin C*?

A) $\frac{1}{\sqrt{2}}$

B) $\frac{\sqrt{3}}{2}$

C) $\frac{1}{2}$

D) 1

279

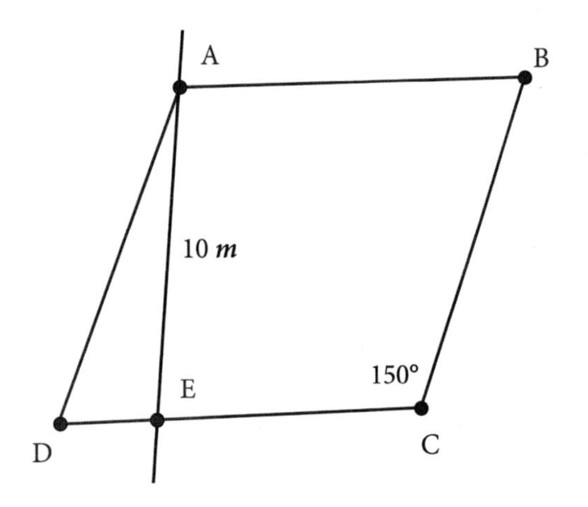

In the above figure, quadrilateral *ABCD* is a parallelogram. *AE* is perpendicular to *CD* and *AE* = 10 meters. What is the length of *BC*?

A) 10 *m*

B) $10\sqrt{3}$ *m*

C) 20 *m*

D) $20\sqrt{3}$ *m*

280

In a right triangle, if $\angle x + \angle y = 90°$ and $\cos x = \frac{5}{12}$, what is the value of *sin y*?

281

Which of the following is NOT equal to $\sin\frac{\pi}{3}$?

A) $\sin\frac{2\pi}{3}$

B) $\cos\frac{\pi}{6}$

C) $\cos\frac{11\pi}{6}$

D) $\sin\frac{4\pi}{3}$

282

In a right triangle, one angle measures $x°$ where $\tan x° = 1$. What is $\sin(90 - x°)$ to the nearest thousandth?

283

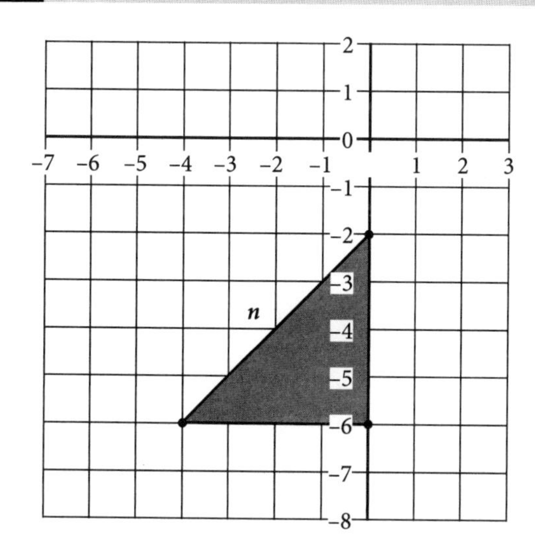

In the graph above, a right triangle is formed with one leg tangent to the y-axis. What is the length of n which is the hypotenuse of the triangle?

A) 2

B) $2\sqrt{2}$

C) 4

D) $4\sqrt{2}$

284

If $\sin a = \cos 26$, what is the value of a?

A) 64

B) 154

C) 206

D) 334

285

If $\sin a = 0.6$ and $\sin (90 - a) = 0.3$, what is the value of $\tan a$?

286

AP is a line tangent to the circle below. If $AO =$ 6 cm and the length of arc length AB is 2π, what is the length of OP in cm?

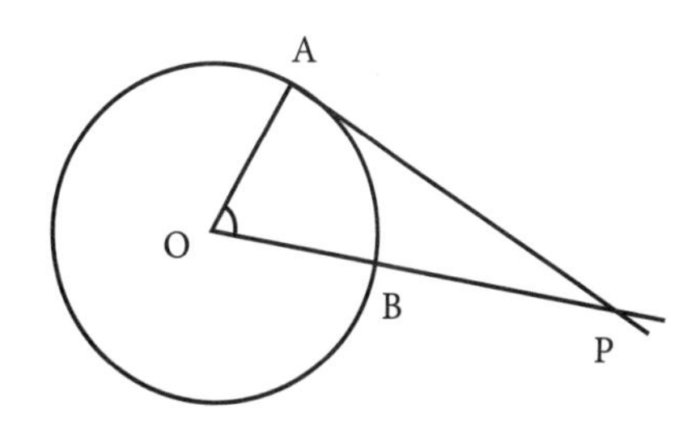

A) $6\sqrt{2}$ cm

B) $6\sqrt{3}$ cm

C) 12 cm

D) 13 cm

287

Triangle PQR and triangle DEF are right-angled triangles. If DE corresponds to PQ, which of the following statements would make the two triangles similar?

A) The length of line DF is twice the length of line PR.

B) The angle $FDE = 60°$.

C) The sin of angle QRP is equal to the sin of angle EFD.

D) The length of the line PR is twice the length of the line DF.

288

In the diagram shown below, PST and PQR are right angles and ST is parallel to QR. What is the value of $\cos R$?

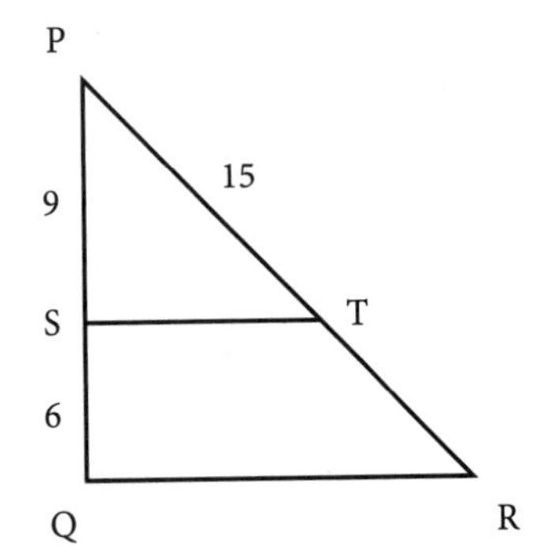

A) $\frac{3}{5}$

B) $\frac{3}{4}$

C) $\frac{4}{5}$

D) $\frac{8}{5}$

289

Triangle DSE has angles $r°$, $p°$ and $90°$. What is the value of $\cos r° - \sin p°$?

A) 0

B) 1

C) $\sqrt{2}$

D) $\sqrt{3}$

278. **Level:** Easy | **Skill/Knowledge:** Right triangles and trigonometry | **Testing Point:** Using concepts and properties of trigonometric ratios and identities to solve problems

Key Explanation: Choice C is correct. Since $\cos A = \frac{1}{2}$, then $A = 60°$.

That means $C = 30°$ (since $B = 90°$).

Therefore, $\sin C = \sin 30° = \frac{1}{2}$.

Distractor Explanations: Choice A is incorrect and may be due to conceptual or calculation error. **Choice B** is incorrect and may result from solving the value of $\sin A$ or $\cos C$. **Choice D** is incorrect and may result from solving the value of $\sin B$.

279. **Level:** Hard | **Skill/Knowledge:** Right triangles and trigonometry | **Testing Point:** Using concepts of supplementary angles, angle sum property, and special triangles to solve problems

Key Explanation: Choice C is correct.
Since adjacent angles of a parallelogram are supplementary, then $\angle D = 180° - \angle C$. Substituting the value of $\angle C$ yields $\angle D = 180° - 150° = 30°$. Since $AE \perp CD$, then $\angle AED = 90°$. Applying the angle sum property of a triangle to find the value of $\angle DAE$ yields $\angle DAE = 180° - 30° - 90° = 60°$. Therefore, ΔAED is a special right triangle (30 – 60 – 90).

We know that the ratio of the sides will be $1:\sqrt{3}:2$.

Given the length of $AE = 10$ meters, then the length of $AD = 2$ times the length of AE which is $2 \times 10 = 20$.

Since the length of opposite sides of a parallelogram are equal, then $AD = BC$.

Therefore, $BC = 20\ m$.

Distractor Explanations: Choice A is incorrect because it is the length of AE. **Choice B** is incorrect because it is the length of DE. **Choice D** is incorrect and may result from a conceptual or calculation error.

280. **Level:** Medium | **Skill/Knowledge:** Right triangles and trigonometry | **Testing Point:** Using definition of trigonometric identities to solve problems

Key Explanation: If two angles are complementary, the cosine of one is equal to the sine of the other. So, $\sin y = \cos x$. Therefore, $\sin y = \frac{5}{12}$.

281. **Level:** Easy | **Skill/Knowledge:** Right triangles and trigonometry | **Testing Point:** Using trigonometric ratios with unit circle to solve problems

Key Explanation: Choice D is correct. Using the unit circle or a calculator (in radians)
$\sin\frac{\pi}{3} = \frac{\sqrt{3}}{2}$ or 0.8660.

Since $\sin\frac{4\pi}{3} = -\frac{\sqrt{3}}{2} \neq \frac{\sqrt{3}}{2}$, **Choice D** is correct.

Distractor Explanations: Choice A is incorrect and is the result of error in interpreting trigonometric ratios on the unit circle. **Choice B** is incorrect and is the result of error in interpreting trigonometric ratios on the unit circle. **Choice C** is incorrect and is the result of error in interpreting trigonometric ratios on the unit circle.

282. **Level:** Medium | **Skill/Knowledge:** Right triangles and trigonometry | **Testing Point:** Using properties of trigonometric ratios and identities to solve problems

Key Explanation: Since $tan\ x° = 1$, then $x° = tan^{-1}(1)$.

Hence, $x° = 45°$.

Therefore, $sin\ (90 - x°) = sin\ (90° - 45°) = sin\ 45°$
$= \frac{1}{\sqrt{2}} = \frac{\sqrt{2}}{2} = \frac{1.414}{2} = 0.707$ (nearest thousandth).

283. **Level:** Easy | **Skill/Knowledge:** Right triangles and trigonometry | **Testing Point:** Using concepts of special right triangles to solve problems

Key Explanation: Choice D is correct. Based on the given graph, the legs of the triangle both measure 4 units. Since the legs are congruent, this is an isosceles right triangle. This means that it is a 45° – 45° – 90° triangle whose sides have a ratio of $x : x : x\sqrt{2}$. Since the lengths of the legs are 4 units, $n = 4\sqrt{2}$.

Distractor Explanations: Choice A is incorrect and reflects error in interpreting angle and side relationship for geometric shapes on a graph. **Choice B** is incorrect and reflects error in interpreting angle and side relationship for geometric shapes on a graph. **Choice C** is incorrect and reflects error in interpreting angle and side relationship for geometric shapes on a graph.

284. **Level:** Easy | **Skill/Knowledge:** Right triangles and trigonometry | **Testing Point:** Using trigonometric identities

Key Explanation: Choice A is correct. Based on the identity $sin\ x = cos\ (90 - x)$, the angles a and 26° are complementary. This means that $a = 90 - 26$. Therefore, $a = 64°$.

Distractor Explanations: Choice B is incorrect and may result from determining the angle supplement to 26°. **Choice C** is incorrect and may be a result of conceptual or calculation error. **Choice D** is incorrect and may result from determining the angle conjugate to 26°.

285. **Level:** Easy | **Skill/Knowledge:** Right triangles and trigonometry | **Testing Point:** Using trigonometric identities to solve problems

Key Explanation: 2 is the correct answer. $tan\ a$ is equivalent to $\frac{sin\ a}{cos\ a}$. Using the identity $sin\ a = cos(90 - a)$, $sin(90 - a)$ is $cos\ a$. Therefore, $tan\ a = \frac{0.6}{0.3} = 2$.

286. **Level:** Medium | **Skill/Knowledge:** Right triangles and trigonometry | **Testing Point:** Using circle proportionality theorems and special triangles

Key Explanation: Choice C is correct. First, find angle AOB using arc length of AB and the radius of the circle. Arc length is found using the formula $\frac{\theta}{360}\pi D$, where D is the diameter of the circle. Since AO is the radius of the circle and $AO = 6$, then $D = 6(2) = 12$. Substituting the diameter to the formula yields $\frac{\angle AOB}{360}\pi(12) = 2\pi$. Multiplying $\frac{360}{12\pi}$ to both sides of the equation yields $\angle AOB = 60°$. Triangle AOP is a right-angled triangle and one of the angle $AOP = 60°$ making the triangle a special 90–60–30 triangle.

Looking at the triangle, *OP* is the hypotenuse. Using the special triangle identity, $OP = 2AO$ or $OP = 2(6\ cm) = 12\ cm$.

Distractor Explanations: Choice A is incorrect and may be due to conceptual or calculation error. **Choice B** is incorrect. This is the value of *AP*. **Choice D** is incorrect and may be due to conceptual or calculation error.

287. **Level:** Easy | **Skill/Knowledge:** Right triangles and trigonometry | **Testing Point:** Working with similar right triangles

Key Explanation: Choice C is correct. In similar triangles, the matching angles are equal. If the sin of angle *QRP* is equal to the sin of angle *EFD*, the angle *QRP* and angle *EFD* are equal.

Distractor Explanations: Choice A is incorrect and does not prove that the two triangles are similar. **Choice B** is incorrect and does not prove that the two triangles are similar. **Choice D** is incorrect and does not prove that the two triangles are similar.

288. **Level:** Easy | **Skill/Knowledge:** Right triangles and trigonometry | **Testing Point:** Using similar triangles and trigonometric ratios to solve problems

Key Explanation: Choice C is correct. Triangle *PQR* is similar to triangle *PST* which means that $\cos PTS = \cos PRQ$. $\cos PTS = \frac{ST}{PT}$. Using Pythagorean theorem $a^2 + b^2 = c^2$ yields $9^2 + ST^2 = 15^2$. Subtracting 9^2 from both sides of the equation yields $ST^2 = 15^2 - 9^2 = 144$. Getting the square root of both sides of the equation yields $ST = 12cm$. Substituting the value of *ST* yields $\cos PTS = \frac{ST}{PT} = \frac{12}{15} = \frac{4}{5}$.

Distractor Explanations: Choice A is incorrect and is the value of *sin PTS*. **Choice B** is incorrect and is the value of *tan PTS*. **Choice D** is incorrect. The value of cosine cannot be greater than 1.

289. **Level:** Easy | **Skill/Knowledge:** Right triangles and trigonometry | **Testing Point:** Using trigonometric identities to solve problems

Key Explanation: Choice A is correct. Using the trigonometric identity $\sin x = \cos(90 - x)$, then *r* and *p* are complementary angles. Therefore, $\cos r° = \sin p°$ which means $\cos r° - \sin p° = 0$.

Distractor Explanations: Choice B is incorrect and may be due to conceptual issue. **Choice C** is incorrect and may be due to conceptual issue. **Choice D** is incorrect and may be due to conceptual issue.

290

If a circle has a center (1, 3) and a radius of $\sqrt{17}$ which of the following points lies on the circle?

A) (1, 5)

B) (0, 7)

C) (7, 0)

D) (7, 1)

291

What is the center of the circle represented by the equation $x^2 - 8x + y^2 - 2y = -8$?

A) (1, 4)

B) (−4, −1)

C) (4, 1)

D) (3, 3)

292

A circle at the origin has a radius of $\sqrt{2}$. Which ordered pair below represents a point on the circle?

A) (2, 0)

B) (1, 1)

C) ($\sqrt{2}$, 1)

D) (0, 0)

293

$$(x - 2)^2 + (y + 1)^2 = 16$$

A student graphed the equation of the circle above on the coordinate plane. Point *O* is the center of the circle and Point *N* is a point on the circle. If segment *ON* is the radius of the circle, which could NOT be a coordinate for Point *N*?

A) (2, 3)

B) (6, −1)

C) (−2, −1)

D) (2, −6)

294

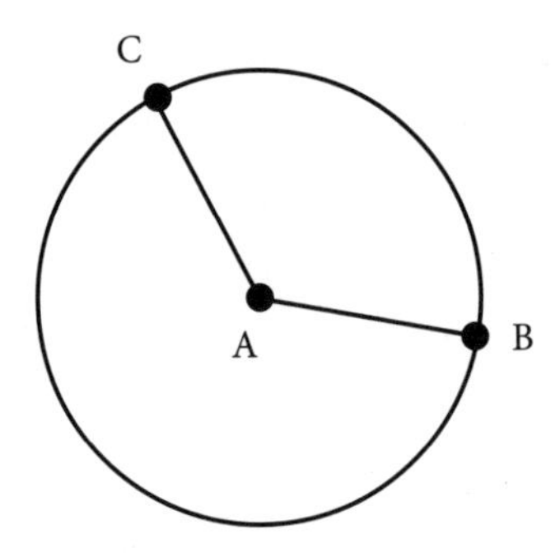

(Note: Figure is not drawn to scale.)

In the circle above, Point *A* is the center and the length of minor arc *BC* is $\frac{3}{8}$ of the length of the circumference of the circle. What is the difference in angles (in degrees) subtended by minor arc *BC* and major arc *BC*?

295

If a sector's interior angle is 72° and the radius is 5 cm, what is the area of the sector? Round your answer the nearest tenth.

296

The diameter of a circle is $2x^3$. What is the area of the circle?

A) πx^2

B) πx^3

C) πx^5

D) πx^6

297

If the length of the minor arc subtended by the angle BOC is 11π, what is the circumference of the circle below?

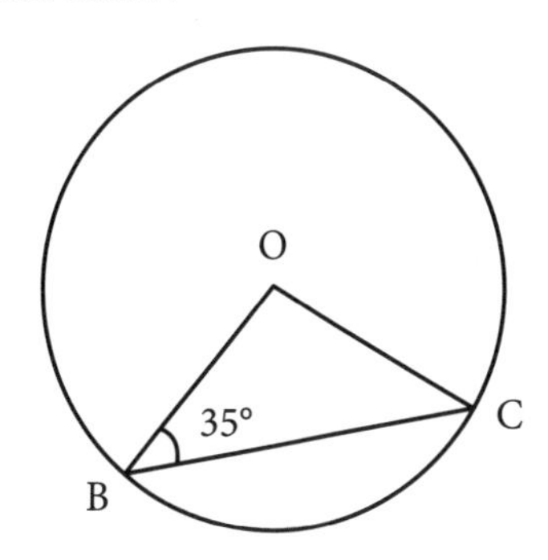

A) 18π

B) 36π

C) 110π

D) 324π

298

PQ is tangent to the circle below and $OP = 12\ cm$. If the arc length subtended by the angle POQ is equal to $\frac{5}{3}\pi$, what is the value of angle PQO in degrees?

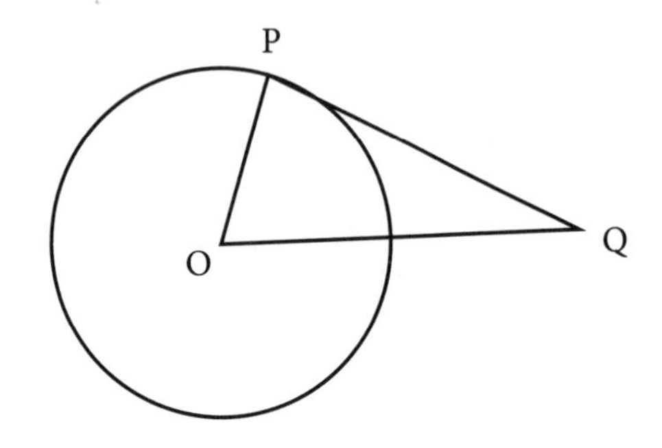

A) 25

B) 30

C) 60

D) 65

299

The circumference of circle A is $2k$ and the circumference of circle B is $36k$. What is the ratio of the area of circle A to circle B?

A) 1 to 36

B) 1 to 18

C) 1 to 324

D) 4 to 324

300

What is the diameter of the circle whose equation is $(x - 3)^2 + (y + 4)^2 = 64$?

290. **Level:** Medium | **Skill/Knowledge:** Circles **Testing Point:** Using definitions, properties, and theorems relating to circles to solve problems

Key Explanation: Choice B is correct. To find an endpoint of a circle, find the equation of a circle from the center and the radius: The equation of a circle in standard form is $(x - h)^2 + (y - k)^2 = r^2$ where (h, k) is the center of the circle and r is the radius.

Substituting the given center and radius yields $(x - 1)^2 + (y - 3)^2 = (\sqrt{17})^2$.

Simplifying the right side of the equation yields $(x - 1)^2 + (y - 3)^2 = 17$.

To identify which points lies on the circle, substitute the values to the equation of the circle.

Substituting (0, 7) from **Choice B** to $(x - 1)^2 + (y - 3)^2 = 17$ yields $(0 - 1)^2 + (7 - 3)^2 = 17$.

Simplifying the equation yields 17=17.

Since the statement is correct, then **Choice B** is the correct answer.

Distractor Explanations: Choices A, C, and **D** are incorrect because substituting these points into the equation of the circle, the results are false statements.

291. **Level:** Hard | **Skill/Knowledge:** Circles | **Testing Point:** Finding the center of a circle using the equation of a circle and completing the square

Key Explanation: Choice C is correct. To find the center of the circle, complete the square and rewrite the expression in the center form: $(x - h)^2 + (y - k)^2 = r^2$.

To determine the number that must be added to complete the squares, divide the coefficient of x and y by 2 and square it. This yields $\left(\frac{-8}{2}\right)^2 = (-4)^2 = 16$ for x and $\left(\frac{-2}{2}\right)^2 = (-1)^2 = 1$ for y. Completing the squares by adding 16 and 1 to both sides of the equation and grouping the squares yields $(x^2 - 8x + 16) + (y^2 - 2y + 1) = -8 + 16 + 1$. Factoring the left side and simplifying the right side of the equation yields $(x - 4)^2 + (y - 1)^2 = 9$. Therefore, the center is (4, 1).

Distractor Explanations: Choice A is incorrect and may result from swapping the values of h and k. **Choice B** is incorrect and may result from negating the values of h and k. **Choice D** is incorrect and may result from solving for the radius and using it as the coordinate of the center.

292. **Level:** Easy | **Skill/Knowledge:** Circles | **Testing Point:** Creating and interpreting the equation of a circle

Key Explanation: Choice B is correct. Writing an equation that represents the given circle with the center at the origin and a radius of $\sqrt{2}$ yields $x^2 + y^2 = 2$. Substituting (1, 1) into the equation yields $1^2 + 1^2 = 2$. Simplifying the equation yields $2 = 2$. Since the statement is true, then (1, 1) lies on the circle.

Distractor Explanations: Choice A is incorrect because substituting (2, 0) yields a false statement of $4 = 2$. **Choice C** is incorrect because substituting ($\sqrt{2}$, 1) yields a false statement of $3 = 2$. **Choice D** is incorrect because (0, 0) is the center of the circle.

293. **Level:** Hard | **Skill/Knowledge:** Circles | **Testing Point:** Using knowledge of equation of circle and radius to solve problem

Key Explanation: Choice D is correct. The equation of a circle is $(x - h)^2 + (y - k)^2 = r^2$, with the center at (h, k) and r being the radius. Therefore, the center of the circle for $(x - 2)^2 + (y + 1)^2 = 16$ is $(2, -1)$ and the radius is 4. A radius of 4 means that a distance of 4 is the farthest any point on the circle can be from the center of the circle. Subtracting 4 from the y-coordinate of the center of the circle gives the point $(2, -5)$ which would be the minimum y value. Since **Choice D** gives a y-coordinate of -6, this point is outside of the circle and cannot be Point N since Point N is on the circle. All of the other answer choices have a y-coordinate within 4 of the center of the circle and therefore cannot be outside of the circle.

Distractor Explanations: Choice A is incorrect and reflects error in understanding the equation of a circle. **Choice B** is incorrect and reflects error in understanding the equation of a circle. **Choice C** is incorrect and reflects error in understanding the equation of a circle.

294. **Level:** Medium | **Skill/Knowledge:** Circles **Testing Point:** Using circle proportionality theorems to find central angles

Key Explanation: The correct answer is 90. Start by finding the degrees of the major and minor arcs. Circle proportionality theorem states that the proportion that an arc length is to the circumference of a circle is the same proportion as the central angle that is subtended by the arc. Thus:

The angle subtended by minor arc $BC = \frac{3}{8} \times 360° = 135°$.

Since minor arc BC and major arc BC comprise the whole circle, the degree measure of major arc BC can be found by subtracting the degree measure of minor arc BC from 360° as follows:

The angle subtended by major arc $BC = 360 - 135 = 225°$.

Therefore, the difference in the angles is $225 - 135 = 90°$.

295. **Level:** Easy | **Skill/Knowledge:** Circles | **Testing Point:** Finding the area of a sector of a circle

Key Explanation: The correct answer is 15.7 cm^2. First, use the degree measure of the interior angle to determine the fraction relationship between the sector and the full circle which yields $\frac{72}{360} = \frac{1}{5}$. Then, find the area of the full circle which is $A = \pi(5)^2 = 25\pi$.

Multiplying the fraction of the sector to the area of the whole circle yields $25\pi \times \frac{1}{5} = 5\pi = 15.7\ cm^2$.

296. **Level:** Easy | **Skill/Knowledge:** Circles | **Testing Point:** Using properties of circles to solve area problems

Key Explanation: Choice D is correct. If the diameter of a circle is $2x^3$, then the radius is $\frac{2x^3}{2} = x^3$.

Therefore, the area of the circle is $\pi\left(x^3\right)^2 = \pi x^6$.

Distractor Explanations: Choice A is incorrect and is the result of error in calculating the area of a circle with algebraic expressions. **Choice B** is incorrect and is the result of error in calculating the area of a circle with algebraic expressions. **Choice C** is incorrect and is the result of error in calculating the area of a circle with algebraic expressions.

297. **Level:** Easy | **Skill/Knowledge:** Circles | **Testing Point:** Using circle theorems

Key Explanation: Choice B is correct. *OB* and *OC* are the radii of the circle and are equal. This makes angles *OBC* and *OCB* equal as they are base angles. The value of angle *BOC* is therefore 180° – 35° – 35° = 110°.

The arc length of the sector can be found by $\frac{\theta}{360}C$ where *C* is the circumference of the circle and θ is the subtended angle. Substituting 110° to θ and equating the formula to the given arc length yields $\frac{110}{360}C = 11\pi$. Multiplying 360 and dividing 110 from both sides of the equation yields $C = 36\pi$.

Distractor Explanations: Choice A is incorrect and may result from solving half of the circumference. **Choice C** is incorrect and may be a result of a conceptual or calculation error. **Choice D** is incorrect and may result from solving the area of the circle.

298. **Level:** Medium | **Skill/Knowledge:** Circles **Testing Point:** Determining the measure of an external angle to a circle

Key Explanation: Choice D is correct. If *PQ* is tangent to the circle, the angle *OPQ* is right-angled. The arc length of the sector can be found by $\frac{\theta}{360}\pi D$ where *D* is the diameter of the circle. Since the $OP = 12$ *cm*, the radius is 12 *cm*. This makes the diameter equal to 24 *cm*. Substituting the diameter and equating the formula to the given arc length yields $\frac{\theta}{360}(\pi)(24) = \frac{5}{3}\pi$. Multiplying both sides of the equation by $\frac{360}{24\pi}$ yields $\theta = 25°$. Angles inside a triangle add up to 180°. Since two of the angles are 90° and 25°. The third angle (*PQO*) is 65°.

Distractor Explanations: Choice A is incorrect. This is the value of angle *POQ*. **Choice B** is incorrect. This option may be a result of conceptual or calculation error. **Choice C** is incorrect. This option may be a result of conceptual or calculation error.

299. **Level:** Medium | **Skill/Knowledge:** Circles **Testing Point:** Working with similar figures

Key Explanation: Choice C is correct. The ratio of the circumferences is $2k{:}36k$ which is 1:18. The ratio of their areas would be the square of the ratio of their circumference. The ratio would therefore be $1^2{:}18^2$ or 1:324.

Distractor Explanations: Choice A is incorrect and may result from calculation or conceptual error. **Choice B** is incorrect. This is the ratio of the circumference. **Choice D** is incorrect and may result from calculation or conceptual error.

300. **Level:** Easy | **Skill/Knowledge:** Circles | **Testing Point:** Using the equation of a circle

Key Explanation: 16 is the correct answer. The standard equation of a circle is given by the equation $(x - h)^2 + (y - k)^2 = r^2$, where *r* is the radius of the circle. Since $r^2 = 64$, then $r = 8$. The diameter of a circle is twice its radius. Therefore, the diameter of the above circle will be 16.

Chapter 7

Math Test

This chapter includes parts:

- Module 1
- Module 2
- Answer Key
- Answers & Explanations

Math

22 QUESTIONS | 35 MINUTES

DIRECTIONS

The questions in this section address a number of important math skills. Use of a calculator is permitted for all questions.

NOTES

Unless otherwise indicated: • All variables and expressions represent real numbers. • Figures provided are drawn to scale. • All figures lie in a plane. • The domain of a given function is the set of all real numbers x for which $f(x)$ is a real number.

REFERENCE

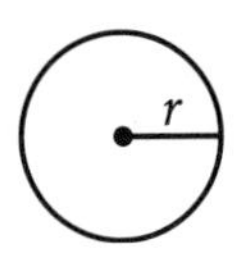

$A = \pi r^2$

$C = 2\pi r$

$A = \ell w$

$A = \frac{1}{2}bh$

$c^2 = a^2 + b^2$

Special Right Triangles

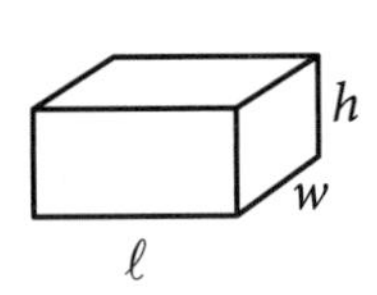

$V = \ell wh$

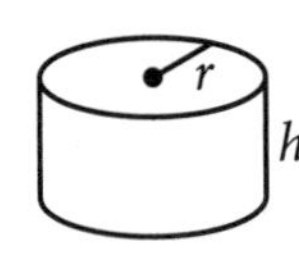

$V = \pi r^2 h$

$V = \frac{4}{3}\pi r^3$

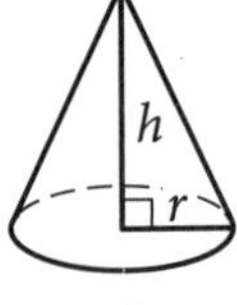

$V = \frac{1}{3}\pi r^2 h$

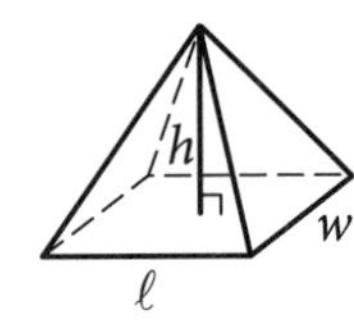

$V = \frac{1}{3}\ell wh$

The number of degrees of arc in a circle is 360.

The number of radians of arc in a circle is 2π.

The sum of the measures in degrees of the angles of a triangle is 180.

For **multiple-choice questions,** solve each problem, choose the correct answer from the choices provided, and then circle your answer in this book. Circle only one answer for each question. If you change your mind, completely erase the circle. You will not get credit for questions with more than one answer circled, or for questions with no answers circled.

For **student-produced response questions,** solve each problem and write your answer next to or under the question in the test book as described below.

- Once you've written your answer, circle it clearly. You will not receive credit for anything written outside the circle, or for any questions with more than one circled answer.
- If you find more than one correct answer, write and circle only one answer.
- Your answer can be up to 5 characters for a positive answer and up to 6 characters (including the negative sign) for a negative answer, but no more.
- If your answer is a fraction that is too long (over 5 characters for positive, 6 characters for negative), write the decimal equivalent.
- If your answer is a decimal that is too long (over 5 characters for positive, 6 characters for negative), truncate it or round at the fourth digit.
- If your answer is a mixed number (such as $3\frac{1}{2}$), write it as an improper fraction (7/2) or its decimal equivalent (3.5).
- Don't include symbols such as a percent sign, comma, or dollar sign in your circled answer.

1

If $f(1) = -4$ and $f(5) = 12$, what is the value of the y-intercept of the linear function $f(x)$?

A) -8

B) 2

C) 4

D) 8

2

The average score in a high school's French entrance exam was 65%. Three new students appeared for the entrance exam and scored 25%, 45%, and 70%; the new average score fell to 60%. How many students had taken the entrance exam before these three students?

A) 4

B) 7

C) 8

D) 11

3

If $sin(4x + 6) = cos(x + 4)$, what is the value of x?

4

$$3y - 2x = 1$$

$$2y + 3x = 18$$

Using the system of equations, what is the value of $5x - y$?

A) -17

B) 4

C) 17

D) 20

5

The function $c(x)$ represents the total cost of a company. The function represents the variable costs and fixed costs of the company. $c(x) = x^2 - 750x + 90,000$, where x represents the number of units produced by the company. How many units would the company have to produce to incur a minimum total cost?

A) 150

B) 375

C) 600

D) 750

6

If $\frac{-3+\sqrt{n}}{-4}$ is a solution to the equation $-2x^2 + 3x + 6 = 0$, what is the value of n?

7

What is the value of $\frac{8}{3}\pi$ radians in degrees?

A) 120°

B) 240°

C) 480°

D) 960°

8

$$|2x - 5| < 3$$

How many integer solutions does the system above have?

A) No solution

B) One solution

C) Two solutions

D) Infinitely many solutions

9

$$3y + 5x = 15$$

What is the x-intercept of the equation of the line shown above?

A) -5

B) -3

C) 3

D) 5

10

The function $f(x) = 3x - 4$ is moved to the left by 4 units and up 2 units. What is the value of its y-intercept?

11

If r and q are solutions to $3x^2 + 6x - 21 = 0$, what is the value of $r + q$?

A) -7

B) -2

C) 2

D) 7

12

Which of the following is equivalent to $7w(px + yz)$?

A) $7wpx + 7wyz$

B) $7wpxyz$

C) $7wpx + yz$

D) $7px + wyz$

13

A cylinder with an open top has a radius of 14 *cm* and a height of 10 *cm*. What is the value of the surface area of the cylinder?

A) 476π

B) 672π

C) $1,400\pi$

D) $1,960\pi$

14

If the expression $\frac{3x^2-6x+11}{x-4}$ is equivalent to $3x+6+\frac{m}{x-4}$, what is the value of m?

15

A study was conducted on students in a summer class in City W. The study showed that 4 in 5 students in the summer class showed an improvement in scores in their opening exam. Which of the following could be concluded from the survey?

A) All the students in City W will improve their entrance scores if they attend the summer class.

B) The students who do not pass the entrance exam should attend the summer class to improve their scores.

C) The summer class will improve the scores of 80% of the students who attend the class.

D) There isn't a cause-effect relationship between the summer class and the entrance exam.

16

A circle has a radius of 6 *cm*. If a sector in the circle has an area of 7.5π, what is the value of the angle that subtends the sector at the center in radians?

A) $\frac{5}{24}\pi$

B) $\frac{5}{12}\pi$

C) $\frac{5}{6}\pi$

D) $\frac{5}{4}\pi$

17

What is the solution of the system below?
$3(x - 5) = 2(x - 2)$

18

If line l (shown below) and line m (not shown) have no solution, which of the following could be the equation of line m?

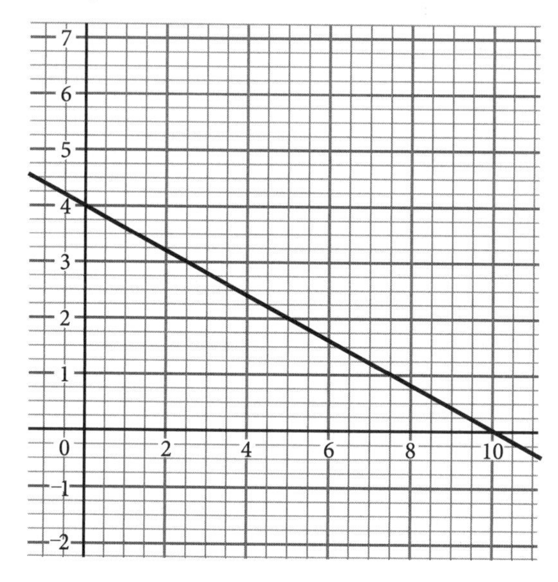

A) $4y - 10x = 6$

B) $10y + 4x = 6$

C) $10x + 4y = 6$

D) $10y - 4x = 6$

19

There are 54 animals (either dogs, cats, or pigs) on a certain Saturday in a park. If there are thrice as many dogs as there are cats and half as many pigs as there are cats on that day. How many dogs were in the park?

A) 6

B) 12

C) 18

D) 36

20

For how many actual points does the line of best fit predict a greater value than the actual value?

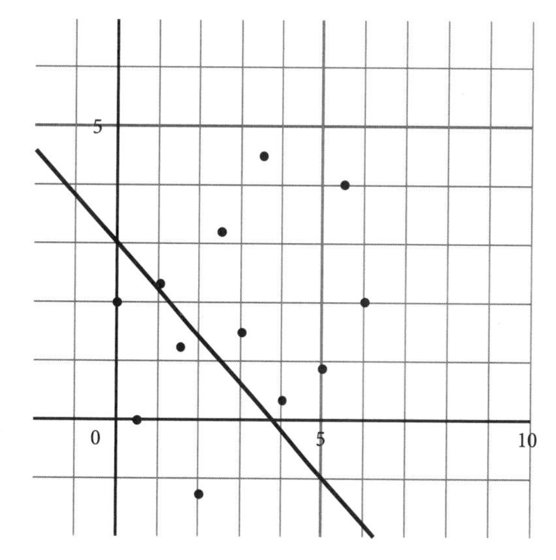

21

$$3y - x = 3$$

$$2y - \frac{4}{3}x = \frac{2}{3}$$

What is the value of x for the given system of equations?

A) -3

B) 1

C) $\frac{5}{3}$

D) 2

22

If $f(2x + 3) = -x^2 + 4x + 6$, what is the value of $f(5)$?

A) 1

B) 6

C) 9

D) 11

No Test Material On This Page

Math

22 QUESTIONS | 35 MINUTES

DIRECTIONS

The questions in this section address a number of important math skills. Use of a calculator is permitted for all questions.

NOTES

Unless otherwise indicated: • All variables and expressions represent real numbers. • Figures provided are drawn to scale. • All figures lie in a plane. • The domain of a given function is the set of all real numbers x for which $f(x)$ is a real number.

REFERENCE

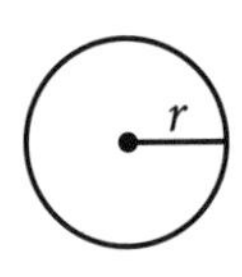

$A = \pi r^2$

$C = 2\pi r$

$A = \ell w$

$A = \frac{1}{2}bh$

$c^2 = a^2 + b^2$

Special Right Triangles

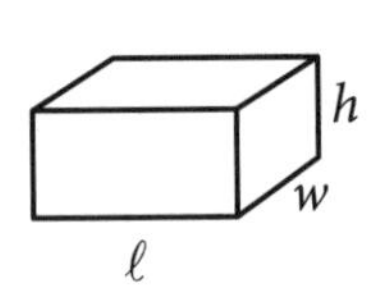

$V = \ell wh$

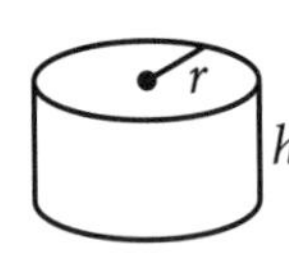

$V = \pi r^2 h$

$V = \frac{4}{3}\pi r^3$

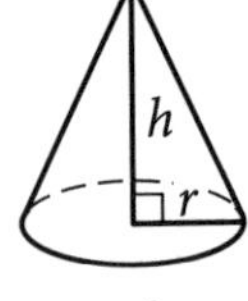

$V = \frac{1}{3}\pi r^2 h$

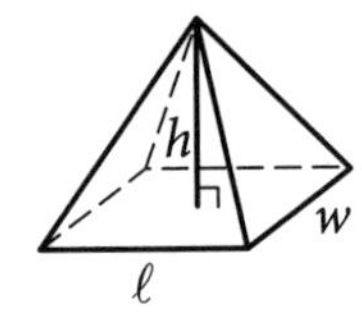

$V = \frac{1}{3}\ell wh$

The number of degrees of arc in a circle is 360.

The number of radians of arc in a circle is 2π.

The sum of the measures in degrees of the angles of a triangle is 180.

For **multiple-choice questions,** solve each problem, choose the correct answer from the choices provided, and then circle your answer in this book. Circle only one answer for each question. If you change your mind, completely erase the circle. You will not get credit for questions with more than one answer circled, or for questions with no answers circled.

For **student-produced response questions,** solve each problem and write your answer next to or under the question in the test book as described below.

- Once you've written your answer, circle it clearly. You will not receive credit for anything written outside the circle, or for any questions with more than one circled answer.
- If you find more than one correct answer, write and circle only one answer.
- Your answer can be up to 5 characters for a positive answer and up to 6 characters (including the negative sign) for a negative answer, but no more.
- If your answer is a fraction that is too long (over 5 characters for positive, 6 characters for negative), write the decimal equivalent.
- If your answer is a decimal that is too long (over 5 characters for positive, 6 characters for negative), truncate it or round at the fourth digit.
- If your answer is a mixed number (such as $3\frac{1}{2}$), write it as an improper fraction (7/2) or its decimal equivalent (3.5).
- Don't include symbols such as a percent sign, comma, or dollar sign in your circled answer.

1 Module 2 1

1

$$p(x) = 60{,}040(1.012)^x$$

The function $p(x)$ models the population of City Y from the year 2004 to the year 2016. Which of the following statements best represents $(1.012)^x$?

A) The population of City Y increases by 12% every year from 2004 to 2016.

B) The population of City Y increases by 1.2% every year from 2004 to 2016.

C) The population of City Y decreases by 12% every year from 2004 to 2016.

D) The population of City Y decreases by 1.2% every year from 2004 to 2016.

2

The number of people who own houses compared to those who rent houses in City L is represented by the ratio 2:3. If the people in City K, with a population of 623,000 are twice as likely to own a house compared to those in City L, then approximately how many people rent houses in City K?

A) 124,600

B) 249,200

C) 373,800

D) 498,400

3

$$4x - 2y = 10$$

$$3y = 2x - 3$$

Using the systems of equation, what is the value of x?

4

Triangle ABC is shown below. If $AB = 36\ cm$ and $AC = 27\ cm$, what is the value of BD?

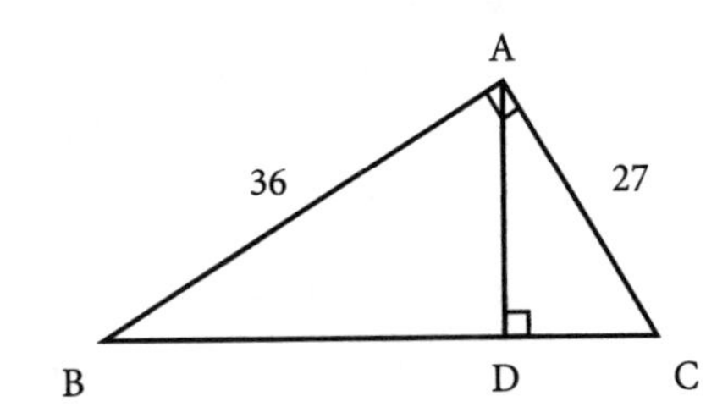

A) 19.29

B) 21.60

C) 25.71

D) 28.80

5

$$y = 2 - k$$

$$y = 3x^2 - 6x + 1$$

If the system of the equation above has two distinct solutions, which of the following cannot be the value of k?

A) 1

B) 2

C) 3

D) 4

6

$$x^2 + 6x + y^2 - 8y = 24$$

What is the value of the x-coordinate of the center of the circle above?

A) −3

B) −4

C) 3

D) 4

7

The length of a rectangular field is 7 meters more than its width. If the area of the field is 120 square meters. What is the value of its length?

8

$$3y > 2x + 5$$

$$y \leq x + 3$$

Which of the following is a solution to the inequality above?

A) (3, 2)

B) (−2, 1)

C) (2, 6)

D) (−1, 4)

9

Which of the following expressions is equivalent to $2x - 6 = 4x + 6$?

A) $\frac{3}{5}x - 12 = 0$

B) $2x = 6$

C) $3x - 18 = 0$

D) $3x + 18 = 0$

10

What is the value of 32% of 500?

11

$$\frac{5}{2x-1}+\frac{3}{2x}$$

Which of the following expressions is equivalent to the expression above?

A) $\frac{16x-3}{4x^2-2x}$

B) $\frac{10x-5}{4x^2-2x}$

C) $\frac{7}{2x-1}$

D) $\frac{8}{4x-1}$

12

x	$f(x)$
1	−2
3	4
5	10

What is the equation of the linear function $f(x)$ as shown above?

A) $f(x) = 3x - 5$

B) $f(x) = -2x$

C) $f(x) = x + 1$

D) $f(x) = 2x$

13

If the system of equations has no solution, what is the value of k?

$$(x+4)^2 = x^2 + 3x + k(x+3)$$

14

Which of the following statements best describes the translation between $f(x) = 2(x+5)^2$ to $g(x) = 2x^2 + 4x + 6$?

A) The graph $f(x)$ is translated 4 units to the right and 4 units up.

B) The graph $f(x)$ is translated 1 unit to the right and 6 units up.

C) The graph $f(x)$ is translated 4 units to the right and 6 units up.

D) The graph $f(x)$ is translated 4 units to the left and 4 units up.

15

If $\frac{3+4i}{2-i} = a+bi$, what is the value of b?

A) $\frac{2}{5}$

B) $\frac{4}{5}$

C) $\frac{7}{5}$

D) $\frac{11}{5}$

16

$$|3x - 5| = 7$$

What is the positive solution to the equation above?

17

A mattress and bed shop tabulates the number of orders it receives in a particular week.

	Mattress	No Mattress
Bed	30	45
No Bed	22	18

What is the probability that a customer buys a mattress given that they do not buy a bed?

A) $\frac{22}{115}$

B) $\frac{40}{115}$

C) $\frac{11}{26}$

D) $\frac{11}{20}$

18

Which of the following system of inequalities represents the graph below?

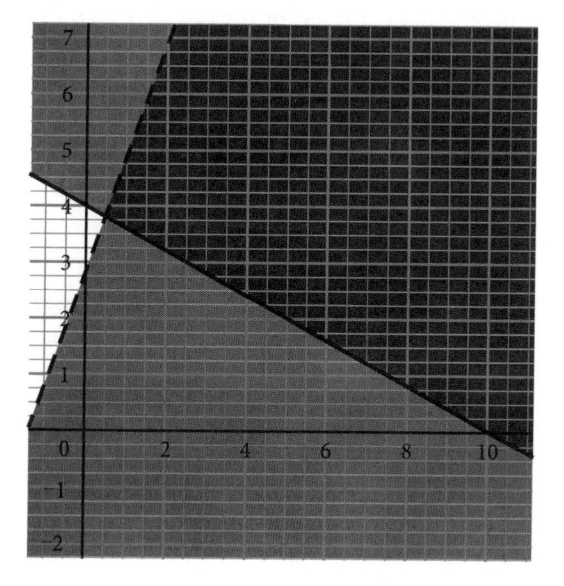

A) $y \geq -0.4x + 4$
$y > 3x + 3$

B) $y \leq 0.4x + 4$
$y < 3x + 3$

C) $y \geq -0.4x + 4$
$y < 3x + 3$

D) $y \leq 0.4x + 4$
$y > 3x + 3$

19

If $a = 2b$ and $\frac{\sqrt[6]{a^4}}{\sqrt[5]{2b}} = a^b$, what is the value of b?

20

Maria traveled at 60 miles per hour for the first two hours of the trip. She then traveled at a speed of 50 miles per hour for the rest of the trip. The average speed of the trip is 54 miles per hour. How long was the whole trip in hours?

A) 2

B) 3

C) 4

D) 5

21

How many solutions does the equation of the graph below have with the line $y = 5$?

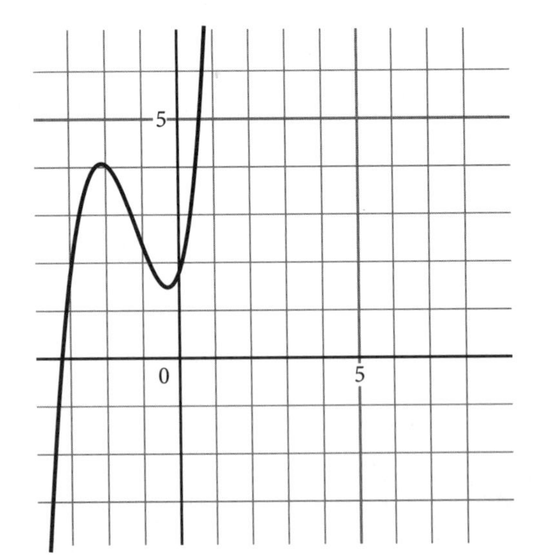

A) None

B) One

C) Two

D) Infinitely many solutions

22

David can spend at most $36 in a bookstore. A book costs $5 and a pen costs $1.50. If he buys at least one book and three pens, what is the maximum number of books that he can get?

A) 4

B) 5

C) 6

D) 7

No Test Material On This Page

Answer Key

Math

Module 1

Questions	Correct	Mark your correct answers
1	A	
2	C	
3	16	
4	C	
5	B	
6	57	
7	C	
8	C	
9	C	
10	10	
11	B	
12	A	
13	A	
14	35	
15	C	
16	B	
17	11	
18	B	
19	D	
20	4	
21	D	
22	C	

Module 2

Questions	Correct	Mark your correct answers
1	B	
2	A	
3	3	
4	D	
5	D	
6	A	
7	15	
8	B	
9	D	
10	160	
11	A	
12	A	
13	5	
14	A	
15	D	
16	4	
17	D	
18	C	
19	$\frac{7}{15}$	
20	D	
21	B	
22	C	

1. **Level:** Easy | **Domain:** ALGEBRA
Skill/Knowledge: Linear functions | **Testing Point:** Solving for the y-intercept of the linear function

Key Explanation: Choice A is correct. The y-intercept is the value of y when x is 0. First, find the equation of the line in slope-intercept form of $y = mx + b$ where m is the slope and b is the y-intercept. Given the points (1, –4) and (5, 12), the slope of the line is $\frac{12-(-4)}{5-1} = \frac{16}{4} = 4$.
Determining the equation of the line using $(y - y_1) = m(x - x_1)$ yields $(y - 12) = 4(x - 5)$.
Using distributive property on the right side of the equation yields $y - 12 = 4x - 20$.
Adding 12 to both sides of the equation yields $y = 4x - 8$.
The equation is $f(x) = 4x - 8$. The y-intercept is therefore –8.

Distractor Explanations: Choice B is incorrect. This is the x-intercept of the function $f(x)$. **Choice C** is incorrect. This is the slope of the function $f(x)$. **Choice D** is incorrect. This is the negative value of y-intercept.

2. **Level:** Medium | **Domain:** PROBLEM-SOLVING AND DATA ANALYSIS
Skill/Knowledge: One-variable data: distributions and measures of center and spread
Testing Point: Mean of data set

Key Explanation: Choice C is correct. Let x be the initial number of students and T be the total score of x number of students.
Since the average score of x students is 65, then $\frac{T}{x} = 65$. Multiplying both sides of the equation by x yields $T = 65x$.
Since 3 students were added, there are $x + 3$ students now.
Adding the scores of the new students yields a new total of $T + 25 + 45 + 70$ or $T + 140$.
Since the new average score is 60, then $\frac{T+140}{x+3} = 60$. Multiplying both sides of the equation by $x + 3$ yields $T + 140 = 60(x + 3)$ or $T + 140 = 60x + 180$.
Since $T = 65x$, then $65x + 140 = 60x + 180$.
Subtracting $60x$ and 140 from both sides of the equation yields $5x = 40$.
Dividing both sides of the equation by 5 yields $x = 8$.
Therefore, there are 8 students initially.

Distractor Explanations: Choice A is incorrect and may be due to calculation or conceptual error. **Choice B** is incorrect and may be due to calculation or conceptual error. **Choice D** is incorrect. This is the total number of students who took the entrance exam, including the three new students.

3. **Level:** Easy | **Domain:** GEOMETRY AND TRIGONOMETRY
Skill/Knowledge: Right triangles and trigonometry | **Testing Point:** Trigonometric identity

Key Explanation: 16 is the correct answer. The trigonometric identity, $sin(x) = cos(90 - x)$, indicates that the angles are complementary. This means that the sum of $4x + 6$ and $x + 4$ is equal to 90. Therefore, $4x + 6 + x + 4 = 90$.
Combining like terms yields $5x + 10 = 90$.
Subtracting 10 from both sides of the equation yields $5x = 80$.
Dividing both sides of the equation by 5 yields $x = 16$.

4. **Level:** Medium | **Domain:** ALGEBRA
Skill/Knowledge: Systems of two linear equations in two variables | **Testing Point:** Solving for linear systems simultaneously

Key Explanation: Choice C is correct. To find the value of $5x - y$, we can subtract the 1st equation from the 2nd equation. This yields $2y + 3x - 3y + 2x = 18 - 1$.
Combining like terms yields $5x - y = 17$.
Therefore, the answer is 17.

Distractor Explanations: Choice A is incorrect and may result from a conceptual or calculation error. **Choice B** is incorrect and may result from a conceptual or calculation error. **Choice D** is incorrect and may result from a conceptual or calculation error.

5. **Level:** Easy | **Domain:** ADVANCED MATH
Skill/Knowledge: Nonlinear functions | **Testing Point:** Determining the vertex of a quadratic function

Key Explanation: Choice B is correct. The number of units that would represent the minimum total cost is the x-coordinate of the vertex of the given function. The x-coordinate of the vertex can be solved using the formula $\frac{-b}{2a}$ where a and b are the coefficients of x^2 and x, respectively. Substituting the values of a and b yields $\left(\frac{-b}{2a}\right)=\left(-\frac{-750}{2}\right)=375$ units.

Distractor Explanations: Choice A is incorrect and may be inferred due to conceptual or calculation error. **Choice C** is incorrect and may be inferred due to conceptual or calculation error. **Choice D** is incorrect and may be inferred due to conceptual or calculation error.

6. **Level:** Medium | **Domain:** ADVANCED MATH
Skill/Knowledge: Nonlinear equations in one variable and systems of equations in two variables | **Testing Point:** Solving for quadratic equations using the quadratic formula

Key Explanation: To solve for this quadratic equation, use the quadratic formula $x=\frac{-b\pm\sqrt{b^2-4ac}}{2a}$, where $a = -2$, $b = 3$ and $c = 6$.
Substituting the values yields $\frac{-3\pm\sqrt{3^2-4(-2)(6)}}{2(-2)}=\frac{3\pm\sqrt{57}}{4}$. Therefore the value of $n = 57$.

7. **Level:** Easy | **Domain:** GEOMETRY AND TRIGONOMETRY
Skill/Knowledge: Lines, angles, and triangles
Testing Point: Converting radians to degrees

Key Explanation: Choice C is correct. π radian is equal to 180°. Therefore, to convert radians to degrees, substitute π with 180° which yields $\frac{8}{3}\times 180° = 480°$.

Distractor Explanations: Choice A is incorrect and may be due to conceptual or calculation error. **Choice B** is incorrect and may be due to conceptual or calculation error. **Choice D** is incorrect and may be due to conceptual or calculation error.

8. **Level:** Medium | **Domain:** ADVANCED MATH
Skill/Knowledge: Nonlinear equations in one variable and systems of equations in two variables | **Testing Point:** Solving for absolute values

Key Explanation: Choice C is correct. To solve for this inequality, the content of the absolute value must be multiplied by 1 and –1 which yields:

$+(2x - 5) < 3$	$-(2x - 5) < 3$
$2x - 5 < 3$	$-2x + 5 < 3$
$2x < 8$	$-2x < -2$
$x < 4.$	$x > 1.$

The integer solutions that fit the inequality $1 < x < 4$ are 2 and 3. Therefore, there are two solutions.

Distractor Explanations: Choice A is incorrect and may be due to conceptual or calculation error. **Choice B** is incorrect and may be due to conceptual or calculation error. **Choice D** is incorrect and may be due to conceptual or calculation error.

9. **Level:** Easy | **Domain:** ALGEBRA
Skill/Knowledge: Linear equations in two variables | **Testing Point:** Finding the x-intercept of the line

Key Explanation: Choice C is correct. The x-intercept of a line is found when the value of y is equal to 0. Substituting 0 to y yields $3(0) + 5x = 15$ or $5x = 15$.
Dividing both sides of the equation by 5 yields $x = 3$.

Distractor Explanations: Choice A is incorrect and may be due to conceptual or calculation error. **Choice B** is incorrect and may be due to conceptual or calculation error. **Choice D** is incorrect and is the value of the y-intercept of the line.

10. **Level:** Medium | **Domain:** ALGEBRA
Skill/Knowledge: Linear functions | **Testing Point:** Linear translation

Key Explanation: To move the function $f(x)$ to the left by 4 units and 2 units up, add 4 to x and add 2 to the constant which yields $f(x) = 3(x + 4) - 4 + 2$.
Using distributive property yields $f(x) = 3x + 12 - 2$.
Simplifying the equation yields $f(x) = 3x + 10$.
The y-intercept is, therefore, 10.

11. **Level:** Easy | **Domain:** ADVANCED MATH
Skill/Knowledge: Nonlinear equations in one variable and systems of equations in two variables | **Testing Point:** Sum of the solutions/roots

Key Explanation: Choice B is correct. In a quadratic equation, the sum of the solutions/roots is given by $\frac{-b}{a}$ where a and b are the coefficients of x^2 and x, respectively. Referring to the given equation, the values of a and b are $a = 3$ and $b = 6$. Therefore, the sum of the solutions will be $\frac{-6}{3} = -2$.

Distractor Explanations: Choice A is incorrect. This option represents the value of the product of the solutions. **Choice C** is incorrect and may be due to conceptual or calculation error. **Choice D** is incorrect and may be due to conceptual or calculation error.

12. **Level:** Easy | **Domain:** ADVANCED MATH
Skill/Knowledge: Equivalent expressions
Testing Point: Distributive property of multiplication

Key Explanation: Choice A is correct. Using the distributive property yields $7w(px + yz) = 7wpx + 7wyz$. This is **Choice A**.

Distractor Explanations: Choice B is incorrect and may be due to conceptual error. **Choice C** is incorrect and may be due to conceptual error. **Choice D** is incorrect and may be due to conceptual error.

13. **Level:** Easy | **Domain:** GEOMETRY AND TRIGONOMETRY
Skill/Knowledge: Area and volume | **Testing Point:** Surface area of an open cylinder

Key Explanation: Choice A is correct. The lateral area of a cylinder can be found by finding the circumference of the base circle and multiplying it by the height of the cylinder which yields ($\pi D \times$ *height of the cylinder*) $= \pi(28) \times 10 = 280\pi$. Since the cylinder has an open top, add the area of the circle at the bottom of the cylinder to find its surface area. The area of the base circle is $\pi r^2 = \pi(14)^2 = 196\pi$. The total surface area of this open-top cylinder would be $196\pi + 280\pi = 476\pi$.

Distractor Explanations: Choice B is incorrect and would give the surface area of the closed cylinder. **Choice C** is incorrect and may be due to calculation or conceptual error. **Choice D** is incorrect. This option gives the volume of the cylinder.

14. **Level:** Easy | **Domain:** ADVANCED MATH
Skill/Knowledge: Equivalent expressions
Testing Point: Remainder theorem

Key Explanation: The variable m is the remainder when $3x^2 - 6x + 11$ is divided by $x - 4$. It can be found by substituting 4, which is the value of x when the divisor is equated to 0(that is, $x - 4 = 0$), into the dividend.
This yields $3(4)^2 - 6(4) + 11 = 35$.

15. **Level:** Easy | **Domain:** PROBLEM-SOLVING AND DATA ANALYSIS
Skill/Knowledge: Evaluating statistical claims: observational studies and experiments | **Testing Point:** Statistical claims

Key Explanation: Choice C is correct. This option is true as 4 out of 5 students who attended the summer class performed better in the opening exam. This is equivalent to 80%.

Distractor Explanations: Choice A is incorrect. Not all students improve their scores after taking the summer class. **Choice B** is incorrect. It is not guaranteed that student scores will improve after attending the summer class. **Choice D** is incorrect. There is a cause-effect relationship between students who attend the summer class and improvement in their entrance exam scores.

16. **Level:** Hard | **Domain:** GEOMETRY AND TRIGONOMETRY
Skill/Knowledge: Lines, angles and triangles
Testing Point: Sector of a circle

Key Explanation: Choice B is correct. To find the proportion of the sector of the circle, first, find the area of the circle which is 36π. The proportion would be $\frac{7.5\pi}{36\pi} = \frac{5}{24}$. Therefore, the angle that would be subtended at the center will be $\frac{5}{24} \times 360 = 75°$. Converting 75° to radians yields $\frac{75}{180} = \frac{5}{12}\pi$.

Distractor Explanations: Choice A is incorrect

because this angle in degrees will be 37.5°. **Choice C** is incorrect because this angle in degrees will be 150°. **Choice D** is incorrect because this angle in degrees will be 135°.

17. **Level:** Easy | **Domain:** ALGEBRA
Skill/Knowledge: Linear equations in one variable | **Testing Point:** System of equations

Key Explanation: Using the distributive property, you get $3x - 15 = 2x - 4$.
Subtracting $2x$ and adding 15 to both sides of the equation yields $x = 11$.
Therefore, the value of x is 11.

18. **Level:** Medium | **Domain:** ALGEBRA
Skill/Knowledge: Linear equations in two variables | **Testing Point:** Parallel lines

Key Explanation: Choice B is correct. For a system of equation to have no solution, the two lines must be parallel. Using the points (0, 4) and (10, 0), the slope of the given line is $\frac{4-0}{0-10} = \frac{4}{-10} = -0.4$. Line m would also, therefore, have a slope of –0.4. Only $10y + 4x = 6$ has a slope of –0.4 and would therefore be the correct answer.

Distractor Explanations: Choice A is incorrect. This line has a slope of $\frac{5}{2}$, the system would be perpendicular and would have one solution. **Choice C** is incorrect and may result from a conceptual or calculation error. **Choice D** is incorrect and may result from a conceptual or calculation error.

19. **Level:** Medium | **Domain:** ALGEBRA
Skill/Knowledge: Linear equations in two variables | **Testing Point:** Linear word problems

Key Explanation: Choice D is correct. Given that c represents the number of cats, the number of dogs d would be $d = 3c$. The number of pigs p would be $p = \frac{1}{2}c$. Therefore, the total number of animals is represented by $c + p + d = 54$ or $c + 3c + \frac{1}{2}c = 54$.
Simplifying the equation yields $\frac{9}{2}c = 54$.
Multiplying $\frac{2}{9}$ to both sides of the equation yields $c = 12$.
Hence, there were 12 cats.
Therefore, the number of dogs is $d = (12 \times 3) = 36$.

Distractor Explanations: Choice A is incorrect. This is the number of pigs that were in the park that day. **Choice B** is incorrect. This is the number of cats that were in the park that day. **Choice C** is incorrect. This is the total number of cats and pigs in the park that day.

20. **Level:** Easy | **Domain:** PROBLEM-SOLVING AND DATA ANALYSIS
Skill/Knowledge: Two-variable data: models and scatterplots | **Testing Point:** Solving for absolute values

Key Explanation: Count the points where the actual value is lower than the value predicted by the line of best fit. These are the points below the line. Therefore, there are 4 points.

21. **Level:** Medium | **Domain:** ALGEBRA
Skill/Knowledge: Linear equations in two variables | **Testing Point:** Solving for linear systems simultaneously

Key Explanation: Choice D is correct. Solve the equation simultaneously. Multiply the first equation by 2 to get $6y - 2x = 6$ and the second equation by 3 to get $6y - 4x = 2$.
Subtract the second equation from the first equation which yields $2x = 4$. Dividing both sides of the equation by 2 yields $x = 2$.

Distractor Explanations: Choice A is incorrect and may be due to conceptual or calculation error. **Choice B** is incorrect and may be due to conceptual or calculation error. **Choice C** is incorrect and is the value of y.

22. **Level:** Hard | **Domain:** ADVANCED MATH
Skill/Knowledge: Nonlinear equations in one variable and systems of equations in two variables | **Testing Point:** Evaluating quadratic equation

Key Explanation: Choice C is correct. To solve for $f(5)$, solve for x by equating $2x + 3$ to 5. This yields $2x + 3 = 5$.
Subtracting 3 from both sides of the equation yields $2x = 2$.
Dividing both sides of the equation by 2 yields $x = 1$.
Substituting 1 for x in the given equation yields $f(2(1) + 3) = -(1)^2 + 4(1) + 6 = 9$.
Therefore, $f(5) = 9$.

Distractor Explanations: Choice A is incorrect. This would be the value of the function if you substitute 5 directly into x. **Choice B** is incorrect and may be due to conceptual or calculation error. **Choice D** is incorrect and may be due to conceptual or calculation error.

1. **Level:** Easy | **Domain:** ADVANCED MATH **Skill/Knowledge:** Nonlinear functions | **Testing Point:** Interpreting exponential functions

   Key Explanation: Choice B is correct. Exponential functions have a format of $y = a(1 + r)^x$ where a is the initial number, $(1 + r)$ is the growth factor, r is the rate of growth and x is the time. Hence, $(1.012)^x$ implies that the population of the town is increasing by a growth factor of 1.012. The growth rate is then 0.012 or 1.2%. **Choice B** is then correct.

   Distractor Explanations: Choice A is incorrect because the statement is false and may be due to conceptual or calculation error. **Choice C** is incorrect because the statement is false and may be due to conceptual or calculation error. **Choice D** is incorrect because the statement is false and may be due to conceptual or calculation error.

2. **Level:** Easy | **Domain:** PROBLEM-SOLVING AND DATA ANALYSIS **Skill/Knowledge:** Ratios, rates, proportional relationships, and units | **Testing Point:** Using ratio in solving problems

   Key Explanation: Choice A is correct. The proportion or percentage of people who own houses in City L is represented by $\frac{2}{5}$ or 40% of the population.

   Since people in City K are twice as likely to own a house, the proportion of people who own a house in City K is $\frac{4}{5}$ or 80% of its population.

   This would imply that only $\frac{1}{5}$ of the population in City K rent houses. This would approximately be $\frac{1}{5} \times 623{,}000 = 124{,}600$ people in City K.

   Distractor Explanations: Choice B is incorrect and would be due to miscalculation and conceptual error. **Choice C** is incorrect and would be due to miscalculation and conceptual error. **Choice D** is incorrect and represents the number of people who own houses in City K.

3. **Level:** Easy | **Domain:** ALGEBRA **Skill/Knowledge:** Systems of two linear equations in two variables | **Testing Point:** Solving for linear systems simultaneously

   Key Explanation: To solve for x, solve for the equations simultaneously. Re-arrange the second equation to get $3y - 2x = -3$.
   Multiply the second equation by 2 which yields $6y - 4x = -6$.
   Add the two equations to get $4y = 4$.
   Dividing both sides of the equation by 4 yields $y = 1$.
   Substitute the y value into the second equation yields $3(1) = 2x - 3$.
   Adding 3 to both sides of the equation yields $6 = 2x$.
   Dividing both sides of the equation by 2 yields $3 = x$ or $x = 3$.

4. **Level:** Medium | **Domain:** GEOMETRY AND TRIGONOMETRY **Skill/Knowledge:** Right triangles and trigonometry | **Testing Point:** Proportions of a triangle

   Key Explanation: Choice D is correct. First find the value of BC, which is the hypotenuse of the triangle ABC. The length of the hypotenuse can be found using the Pythagoras' theorem $c = \sqrt{a^2 + b^2}$, where a and b are 36 and 27 and c is BC. The value of c(the hypotenuse) will be $c = 45$. Using proportions, the lengths of triangle ABC is directly proportional to the lengths of triangle *ABD*. Therefore, $\frac{45}{36}=\frac{36}{BD}$.

$BD = 28.80$.

Distractor Explanations: Choice A is incorrect and may be due to miscalculation and conceptual error. **Choice B** is incorrect. This option is the value of *AD*. **Choice C** is incorrect and may be due to miscalculation and conceptual error.

5. **Level:** Medium | **Domain:** ADVANCED MATH
Skill/Knowledge: Nonlinear equations in one variable and systems of equations in two variables | **Testing Point:** Systems of linear and quadratic equation

Key Explanation: Choice D is correct. To solve for the system, use substitution method which yields $2 - k = 3x^2 - 6x + 1$. Subtracting 2 and adding k to both sides of the equation would result in a quadratic equation: $0 = 3x^2 - 6x + k - 1$. For quadratic equations with two solutions, the discriminant is greater than 0 ($b^2 - 4ac > 0$). In the quadratic equation, $a = 3$, $b = -6$ and $c = k - 1$. Therefore, $(-6)^2 - 4(3)(k - 1) > 0$ which would translate to $36 - 12k + 12 > 0$. Combining the constants yields $48 - 12k > 0$. Subtracting 48 from both sides of the inequality yields $-12k > -48$. Dividing both sides of the inequality by -12 yields $k < 4$. Only **Choice D** is not less than 4. Therefore, it is the correct answer.

Distractor Explanations: Choice A is incorrect. This option would result in two distinct solutions in the quadratic and would therefore be incorrect. **Choice B** is incorrect. This option would result in two distinct solutions in the quadratic and would therefore be incorrect. **Choice C** is incorrect. This option would result in two distinct solutions in the quadratic and would therefore be incorrect.

6. **Level:** Easy | **Domain:** GEOMETRY AND TRIGONOMETRY
Skill/Knowledge: Circles | **Testing Point:** Equations of a circle

Key Explanation: Choice A is correct. The standard equation of a circle is given by $(x - h)^2 + (y - k)^2 = r^2$ where (h, k) is the center of the circle. To convert our current equation to a standard equation, complete the perfect square trinomials in $(x^2 + 6x) + (y^2 - 8y) = 24$.
Adding 9 and 16 to both sides of the equation yields $(x^2 + 6x + 9) + (y^2 - 8y + 16) = 24 + 16 + 9$.
Factoring the perfect square trinomials yields $(x + 3)^2 + (y - 4)^2 = 49$.
Since $h = -3$ and $k = 4$, then the center of the circle is $(-3, 4)$. The x-coordinate is therefore -3.

Distractor Explanations: Choice B is incorrect and is the negative value of the y-coordinate of the center of the circle. **Choice C** is incorrect and may result from finding the value of $-h$. **Choice D** is incorrect and is the y-coordinate of the center of the circle.

7. **Level:** Medium | **Domain:** GEOMETRY AND TRIGONOMETRY
Skill/Knowledge: Area and volume | **Testing Point:** Area of a rectangle

Key Explanation: 15 is the correct answer. The area of a rectangular field is given by *length* (l) × *width* (w). If the width can be given by w, then the length would then be $w + 7$. Therefore, the area would be $w(w + 7) = 120$ which would result to $w^2 + 7w = 120$.
Subtracting 120 from both sides of the equation yields $w^2 + 7w - 120 = 0$.
Solve the quadratic equation to find the value of w.
Splitting the middle term and grouping the binomials yields $(w^2 + 15w) + (-8w - 120) = 0$.

Factoring out w and -8 yields $w(w + 15) - 8(w + 15) = 0$. Factoring out $(w + 15)$ yields $(w + 15)(w - 8) = 0$. Solving the values of w yields $w = 8$ or -15. Since the width of a rectangle cannot be negative, then the value of the width is 8. Therefore, the length would be $8 + 7 = 15$.

8. **Level:** Easy | **Domain:** ALGEBRA
Skill/Knowledge: Linear inequalities in one or two variables | **Testing Point:** Solving for linear inequalities systems

Key Explanation: Choice B is correct. To solve the system of inequality, substitute the solutions into the given inequalities and verify the statements.
For $(-2, 1)$, the first inequality becomes $3(1) > 2(-2) + 5$ which is equivalent to $3 > 1$. This statement is true. Evaluating the second inequality yields, $1 \leq -2 + 3$ which is equivalent to $1 \leq 1$. This statement is also true. This makes $(-2, 1)$ a solution to the system of inequality.

Distractor Explanations: Choice A is incorrect. Substituting$(3, 2)$ to the 1st inequality yields $6 > 11$ which is a false statement. **Choice C** is incorrect. Substituting $(2, 6)$ to the 2nd inequality yields $6 \leq 5$ which is a false statement. **Choice D** is incorrect. Substituting $(-1, 4)$ to the 2nd inequality yields $4 \leq 2$ which is a false statement.

9. **Level:** Easy | **Domain:** ALGEBRA
Skill/Knowledge: Linear equations in one variable | **Testing Point:** Solving for x

Key Explanation: Choice D is correct. First, solve for x in the equation $2x - 6 = 4x + 6$. Subtracting $4x$ and 6 from both sides of the equation yields $-2x - 12 = 0$.
Dividing both sides of the equation by -2 yields $x + 6 = 0$.
Multiplying both sides of the equation by 3 yields $3x + 18 = 0$ which is **Choice D**.
Therefore, $3x + 18 = 0$ is the correct answer.

Distractor Explanations: Choice A is incorrect. Once solved the value of $x = 20$, which is not equivalent to $x = -6$. **Choice B** is incorrect. Once solved, the value of $x = 3$, which is not equivalent to $x = -6$. **Choice C** is incorrect. Once solved, the value of $x = 6$, which is not equivalent to $x = -6$.

10. **Level:** Easy | **Domain:** PROBLEM-SOLVING AND DATA ANALYSIS
Skill/Knowledge: Percentages | **Testing Point:** Percentage of a value

Key Explanation: 160 is the correct answer. To solve for this, convert the percentage into a fraction that yields $\frac{32}{100}$. Then multiply the fraction by the given amount which yields $\frac{32}{100} \times 500 = 160$.

11. **Level:** Medium | **Domain:** ADVANCED MATH
Skill/Knowledge: Equivalent expressions
Testing Point: Adding rational expressions

Key Explanation: Choice A is correct. To solve this problem, add the two expressions by making their denominators the same.
This yields $\frac{5(2x)+3(2x-1)}{2x(2x-1)}$. Using distributive property yields $\frac{10x+6x-3}{4x^2-2x}$. Combining like terms yields $\frac{16x-3}{4x^2-2x}$. Therefore, the answer is **Choice A**.

Distractor Explanations: Choice B is incorrect and may be due to calculation or conceptual error. **Choice C** is incorrect and may be due to calculation or conceptual error. **Choice D** is incorrect and may be due to calculation or conceptual error.

12. **Level:** Medium | **Domain:** ALGEBRA
Skill/Knowledge: Linear functions | **Testing Point:** Finding the equation of the function

Key Explanation: Choice A is correct. To find the equation of the function $f(x)$, first, find the slope of the function. Using points (3, 4) and (5, 10), the slope is $\frac{y_2 - y_1}{x_2 - x_1} = \frac{10-4}{5-3} = \frac{6}{2} = 3$. The equation of the line can then be found using the formula $(y - y_1) = m(x - x_1)$ which would result in $(y - 10) = 3(x - 5)$.
Using distributive property yields $y - 10 = 3x - 15$.
Adding 10 to both sides of the equation yields $y = 3x - 5$.
Therefore, the answer is $f(x) = 3x - 5$.

Distractor Explanations: Choice B is incorrect and may be due to calculation or conceptual error. **Choice C** is incorrect and may be due to calculation or conceptual error. **Choice D** is incorrect and may be due to calculation or conceptual error.

13. **Level:** Easy | **Domain:** ALGEBRA
Skill/Knowledge: Linear equations in one variable | **Testing Point:** System of equations

Key Explanation: Using distributive property yields $x^2 + 8x + 16 = x^2 + 3x + kx + 3k$. Grouping like terms together and subtracting x^2 from both sides of the equation yields $8x + 16 = 3x + kx + 3k$.
Subtracting $3x$ from both sides of the equation yields $5x + 16 = kx + 3k$.
For a system of equations to have no solution, the gradient on the left side of the equation and the gradient on the right side of the equation are equivalent. Therefore $k = 5$.

14. **Level:** Hard | **Domain:** ADVANCED MATH
Skill/Knowledge: Nonlinear functions | **Testing Point:** Translation of function

Key Explanation: Choice A is correct. To compare quadratic functions, the functions should be in their vertex form which is $y = a(x - h)^2 + k$ where (h, k) is the vertex. In the function $f(x)$, the vertex is $(-5, 0)$. To find the vertex of $g(x)$, convert $g(x)$ to its vertex form.
Factoring out 2 yields $g(x) = 2(x^2 + 2x + 3)$.
Completing the perfect square trinomial by grouping yields $g(x) = 2[(x^2 + 2x + 1) + 2]$.
Factoring the perfect square trinomial yields $g(x) = 2[(x + 1)^2 + 2]$.
Distributing 2 to the constant yields $g(x) = 2(x + 1)^2 + 4$.
Hence, the vertex of $g(x)$ is $(-1, 4)$. Since the x coordinate translates from -5 to -1, the function moves 4 units to the right. Since the y coordinate translates from 0 to 4, the function moves 4 units up.

Distractor Explanations: Choice B is incorrect and may be due to calculation or conceptual error. **Choice C** is incorrect and may be due to calculation or conceptual error. **Choice D** is incorrect and may be due to calculation or conceptual error.

15. **Level:** Easy | **Domain:** ADVANCED MATH
Skill/Knowledge: Equivalent expressions
Testing Point: Complex numbers

Key Explanation: Choice D is correct. To solve this expression, rationalize the expressions by multiplying the numerator and denominator by $2 + i$. This yields $\frac{3+4i}{2-i} \cdot \frac{2+i}{2+i} = \frac{6+8i+3i+4i^2}{2^2 - i^2} = \frac{6+11i-4}{4-(-1)} = \frac{2+11i}{5}$. Splitting the terms yields $\frac{2}{5} + \frac{11}{5}i$. Hence, $a = \frac{2}{5}$ and $b = \frac{11}{5}$.

Distractor Explanations: Choice A is incorrect. This is the value of a. **Choice B** is incorrect and may be due to calculation or conceptual error. **Choice C** is incorrect and may be due to calculation or conceptual error.

16. **Level:** Easy | **Domain:** ADVANCED MATH **Skill/Knowledge:** Nonlinear equations in one variable and systems of equations in two variables | **Testing Point:** Solving for absolute values

Key Explanation: 4 is the correct answer. To solve for this absolute value, create two equations by multiplying the contents of the absolute value with +1 and −1.
The first equation will become $+(3x - 5) = +7$.
Simplifying it yields $3x - 5 = 7$.
Adding 5 to both sides of the equation yields $3x = 12$.
Dividing both sides of the equation by 3 yields $x = 4$.
Finding the second solution yields $-(3x - 5) = 7$.
Simplifying the equation yields $-3x + 5 = 7$.
Subtracting 5 from both sides of the equation yields $-3x = 2$.
Dividing both sides of the equation by −3 yields $x = \frac{-2}{3}$.
Therefore, 4 is the positive solution to the absolute value.

17. **Level:** Easy | **Domain:** PROBLEM-SOLVING AND DATA ANALYSIS **Skill/Knowledge:** Probability and conditional probability | **Testing Point:** Probability

Key Explanation: Choice D is correct. Probability is equivalent to $\frac{\text{Favourable outcomes}}{\text{Total outcomes}}$. Therefore, the total number of outcome is equal to the sum of orders that do not include beds. Hence, $22 + 18 = 40$. The favorable outcomes would be the orders that include a mattress without a bed which is 22. The probability would be equivalent to $\frac{22}{40} = \frac{11}{20}$.

Distractor Explanations: Choice A is incorrect and may be due to calculation error. **Choice B** is incorrect and may be due to calculation error. **Choice C** is incorrect and may be due to calculation error.

18. **Level:** Medium | **Domain:** ALGEBRA **Skill/Knowledge:** Linear inequalities in one or two variables | **Testing Point:** Graphing linear inequalities

Key Explanation: Choice C is correct. To find the correct system of inequalities, focus first on the graph with a solid line. Using points (0, 4) and (10, 0), the slope of this line is $\frac{4-0}{0-10} = -0.4$.
Since **Choices A** and **C** have inequalities with the slope of −0.4, then one of them might be the correct answer.
In the second graph with the dashed line, the shading is below it. Hence, the symbol used is <.
Therefore, **Choice C** is the correct answer.

Distractor Explanations: Choice A is incorrect. The graph of the second inequality is shaded above the line. **Choice B** is incorrect as the equation has positive gradient. **Choice D** is incorrect as the equation has positive gradient.

19. **Level:** Easy | **Domain:** ADVANCED MATH **Skill/Knowledge:** Equivalent expressions **Testing Point:** Exponents

Key Explanation: Since $a = 2b$, then

$$\frac{\sqrt[6]{a^4}}{\sqrt[5]{2b}} = \frac{\sqrt[6]{a^4}}{\sqrt[5]{a}} = \frac{a^{\frac{4}{6}}}{a^{\frac{1}{5}}} = a^{\frac{4}{6}-\frac{1}{5}} = a^{\frac{7}{15}}.$$

Since $a^{\frac{7}{15}} = a^b$, then $b = \frac{7}{15}$.

20. **Level:** Hard | **Domain:** PROBLEM-SOLVING AND DATA ANALYSIS
Skill/Knowledge: Ratios, rates, proportional relationships, and units | **Testing Point:** Average speed

Key Explanation: Choice D is correct. Average speed is given by $\frac{\textit{Total Distance Travelled}}{\textit{Total Time Taken}}$.

The distance in the first two hours is $(60 \times 2) =$

120 miles. The distance covered on the rest of the trip is equivalent to $(50 \times x) = 50x$ miles, where x is the number of hours it took to finish the rest trip. Total distance traveled $= 120 + 50x$ miles. Total time taken $= 2 + x$ hours. Therefore, $\frac{120+50x}{2+x} = 54$.

Multiplying both sides of the equation by $2 + x$ yields $120 + 50x = 108 + 54x$.

Subtracting 108 and $50x$ from both sides of the equation yields $12 = 4x$.

Dividing both sides of the equation by 4 yields $3 = x$ or $x = 3$. Therefore, the total time taken $= 2 + 3 = 5$ hours.

Distractor Explanations: Choice A is incorrect and may be due to calculation or conceptual error. **Choice B** is incorrect and may be due to calculation or conceptual error. **Choice C** is incorrect and may be due to calculation or conceptual error.

21. **Level:** Easy | **Domain:** ADVANCED MATH
Skill/Knowledge: Nonlinear functions | **Testing Point:** Systems of linear and nonlinear functions

Key Explanation: Choice B is correct. The line $y = 5$ is a horizontal line parallel to the x-axis and passes through point (0, 5). The line would only pass once through the graph of the nonlinear equation.

Distractor Explanations: Choice A is incorrect and may be due to conceptual error. **Choice C** is incorrect and may be due to conceptual error. **Choice D** is incorrect and may be due to conceptual error.

22. **Level:** Hard | **Domain:** ALGEBRA
Skill/Knowledge: Linear inequalities in one or two variables | **Testing Point:** Applying linear inequalities to solve word problems

Key Explanation: Choice C is correct. The least amount of pens that David can buy is 3 which cost \$4.50. This would mean that David can spend at most \$36 – \$4.5 = \$31.50 on books. If x is the number of books, then $5x < 31.50$. Dividing both sides of the inequality by 5 yields $x < 6.3$. Therefore, he can get a maximum of 6 books without exceeding his budget.

Distractor Explanations: Choice A is incorrect. David can buy 4 books however this would not be the maximum number of books he can get. **Choice B** is incorrect. David can buy 5 books however this would not be the maximum number of books he can get. **Choice D** is incorrect. This exceeds the maximum number of books David can get by spending above \$36.